Video Notebook

Val Villegas
Southwestern College

Elementary & Intermediate Algebra
Third Edition

Michael Sullivan, III
Joliet Junior College

Katherine Struve
Columbus State Community College

Janet Mazzarella
Southwestern College

PEARSON

Boston   Columbus   Indianapolis   New York   San Francisco   Upper Saddle River
Amsterdam   Cape Town   Dubai   London   Madrid   Milan   Munich   Paris   Montreal   Toronto
Delhi   Mexico City   São Paulo   Sydney   Hong Kong   Seoul   Singapore   Taipei   Tokyo

The author and publisher of this book have used their best efforts in preparing this book. These efforts include the development, research, and testing of the theories and programs to determine their effectiveness. The author and publisher make no warranty of any kind, expressed or implied, with regard to these programs or the documentation contained in this book. The author and publisher shall not be liable in any event for incidental or consequential damages in connection with, or arising out of, the furnishing, performance, or use of these programs.

Reproduced by Pearson from electronic files supplied by the author.

Copyright © 2014, 2010, 2007 Pearson Education, Inc.
Publishing as Pearson, 75 Arlington Street, Boston, MA 02116.

All rights reserved. No part of this publication may be reproduced, stored in a retrieval system, or transmitted, in any form or by any means, electronic, mechanical, photocopying, recording, or otherwise, without the prior written permission of the publisher. Printed in the United States of America.

ISBN-13: 978-0-321-88127-4
ISBN-10: 0-321-88127-3

www.pearsonhighered.com

PEARSON

VIDEO NOTEBOOK

Elementary & Intermediate Algebra, Third Edition

Table of Contents

Chapter 1	1		Chapter 4	119
Section 1.2	1		Section 4.1	119
Section 1.3	10		Section 4.2	127
Section 1.4	16		Section 4.3	132
Section 1.5	23		Section 4.4	137
Section 1.6	30		Section 4.5	141
Section 1.7	34		Section 4.6	145
Section 1.8	40			

Chapter 2	45		Chapter 5	149
Section 2.1	45		Section 5.1	149
Section 2.2	48		Section 5.2	155
Section 2.3	53		Section 5.3	159
Section 2.4	57		Section 5.4	165
Section 2.5	62		Section 5.5	174
Section 2.6	69		Section 5.6	180
Section 2.7	72			
Section 2.8	77			

Chapter 3	85		Chapter 6	185
Section 3.1	85		Section 6.1	185
Section 3.2	89		Section 6.2	193
Section 3.3	96		Section 6.3	199
Section 3.4	102		Section 6.4	205
Section 3.5	107		Section 6.5	208
Section 3.6	110		Section 6.6	213
Section 3.7	114		Section 6.7	220

Chapter 7	**223**		**Chapter 10**	**359**
Section 7.1	223		Section 10.1	359
Section 7.2	227		Section 10.2	365
Section 7.3	231		Section 10.3	372
Section 7.4	235		Section 10.4	374
Section 7.5	241		Section 10.5	384
Section 7.6	245		Section 10.6	389
Section 7.7	253		Section 10.7	393
Section 7.8	259			
Getting Ready for Chapter 8	**265**		**Chapter 11**	**395**
			Section 11.1	395
			Section 11.2	404
			Section 11.3	410
Chapter 8	**269**		Section 11.4	419
Section 8.1	269		Section 11.5	423
Section 8.2	272			
Section 8.3	276		**Chapter 12**	**427**
Section 8.4	284		Section 12.1	427
Section 8.5	290		Section 12.2	429
Section 8.6	296		Section 12.3	432
Section 8.7	302		Section 12.4	438
Section 8.8	309		Section 12.5	444
			Section 12.6	449
Chapter 9	**313**			
Section 9.1	313		**Chapter 13**	**451**
Section 9.2	317		Section 13.1	451
Section 9.3	322		Section 13.2	455
Section 9.4	327		Section 13.3	458
Section 9.5	334		Section 13.4	464
Section 9.6	339			
Section 9.7	342		**Appendix A**	**467**
Section 9.8	346			
Section 9.9	350		**Appendix C**	**471**
			Section C.1	471
			Section C.2	478
			Section C.3	483
			Section C.4	490

Course: Name:
Instructor: Section:

Section 1.2 Video Guide
Fractions, Decimals, and Percents

Objectives:
1. Factor a Number as a Product of Prime Factors
2. Find the Least Common Multiple of Two or More Numbers
3. Write Equivalent Fractions
4. Write a Fraction in Lowest Terms
5. Round Decimals
6. Convert Between Fractions and Decimals
7. Convert Between Percents and Decimals

Section 1.2 – Objective 1: Factor a Number as a Product of Prime Factors
Video Length – 3:40

Definition
A natural number is _____ if its only factors are _____ and _____.

Natural numbers that are not prime are called _____. The number 1 is neither prime nor composite.

In this section, we will work on factoring composite numbers. For example, in $3 \cdot 5 = 15$, the 3 and the 5 are called _____ and the 15 is called the _____.

Prime numbers: _____

1. **Example:** Find the prime factorization of 80.

Final answer: 80 = _____

Note: The instructor writes $2 \cdot 2 \cdot 2 \cdot 2$ as 2^4. If you're not familiar with this notation, don't worry about it yet. Exponential notation will be covered in a later section.

Copyright © 2014 Pearson Education, Inc.

Course: Name:
Instructor: Section:

Section 1.2 – Objective 2: Find the Least Common Multiple of Two or More Numbers
Video Length – 9:39

Before we talk about the definition of the least common multiple, let's just talk about what a multiple actually is.

> **Definition**
> A _____ of a number is the _____ of that number and any
> _____ number.

Examples of multiples of 2:

> **Definition**
> The _____ _____ _____ (_____) of two or more natural
> numbers is the _____ number that is a _____ of each of the numbers.

Let's say we want the least common multiple between 8 and 12:

2. **Example:** Find the LCM of the numbers 18 and 15.

Write the steps in words	Show the steps with math
Step 1	
Step 2	
Step 3	

   Final answer: LCM = _____

3. **Example:** Find the LCM of 24 and 30.

   Final answer: LCM = _____

Be sure to make an effort to be neat an organized when you do math! The more organized you are, the easier it is to keep track of your work and not make any BONEHEAD mistakes!!!

Course: Name:
Instructor: Section:

Section 1.2 – Objective 3: Write Equivalent Fractions
Video Length – 7:05

Consider the fraction

$$\frac{3}{4}$$

So how do we go about obtaining equivalent fractions?

Definition
The _____ _____ _____ (_____) is the least common multiple of the denominators of a group of fractions.

Note: Finding the least common denominator is the exact same thing as finding the least common multiple. Except that we are looking at the denominators of a group of fractions.

We are going to combine two ideas into one – we are going to combine the idea of the LCM with the idea of equivalent fractions.

4. **Example:** Write $\frac{5}{8}$ and $\frac{9}{20}$ as equivalent fractions with the least common denominator.

Write the steps in words	Show the steps with math
Step 1	
Step 2	

Final answer: LCD = _____

Copyright © 2014 Pearson Education, Inc.

Course: Name:
Instructor: Section:

Section 1.2 – Objective 4: Write a Fraction in Lowest Terms
Video Length – 3:02

Definition
A fraction is written in _____ _____ if the numerator and the denominator share no common factor other than 1.

In order to write a fraction in lowest terms, use the "reverse" idea of what is used to create equivalent fractions.

When writing a fraction in lowest terms, we factor the numerator as primes, factor the denominator as primes and then divide out those common factors.

5. **Example:** Write $\dfrac{15}{55}$ in lowest terms.

   Final answer: $\dfrac{15}{55}$ = _____

6. **Example:** Write $\dfrac{16}{84}$ in lowest terms.

   Final answer: $\dfrac{16}{84}$ = _____

Course:
Instructor:

Name:
Section:

Section 1.2 – Objective 5: Round Decimals
Video Length – 3:37

We spent a lot of time talking about fractions, now we will spend a little a bit of time talking about decimals.

$$7 \,.\, 4 \; 6 \; 3 \; 1 \; 5 \; 2$$

For example, 0.3 can be read as "point three" or "three-_____". This is the same as

0.3 = _____ . What does 0.28 mean? 0.28 = _____

Round Decimals
We round decimals in the same way we round whole numbers. First, identify the specified place value in the decimal. If the digit to the _____ is _____ or _____, add _____ to the digit; if the digit to the _____ is _____ or _____, leave the digit as it is. Then drop the digits to the right of the specified place value.

7. **Example:** Round 0.9451 to the nearest thousandth.

   Final answer: _____

8. **Example:** Round 4.359 to the nearest hundredth.

   Final answer: _____

Course:
Instructor:

Name:
Section:

Section 1.2 – Objective 6: Convert Between Fractions and Decimals
Part I – Convert a Decimal to a Fraction
Video Length – 2:13

Convert a Decimal to a Fraction
To convert a decimal to a fraction, identify the place value of the _____ digit in the decimal.

Write the decimal as a _____ using the place value of the last digit as the

_____, and write in lowest terms.

9. **Example:** Convert each decimal to a fraction and write in lowest terms, if possible.

   (a) 0.4

   (b) 0.531

   Final answer: 0.4 = _____

   Final answer: 0.531 = _____

Course:
Instructor:
Name:
Section:

Section 1.2 – Objective 6: Convert Between Fractions and Decimals
Part II – Convert a Fraction to a Decimal
Video Length – 4:38

Convert a Fraction to a Decimal
To convert a fraction to a decimal, _____ the _____ of the fraction by the _____ of the fraction until the _____ is _____ or the _____ _____.

10. **Example:** Convert $\frac{7}{25}$ to a decimal.

    Note: Pay attention to what he says about "terminating".

    Final answer: $\frac{7}{25} =$ _____

11. **Example:** Convert $\frac{7}{9}$ to a decimal.

    Note: Pay attention to the notation that is used.

    Final answer: $\frac{7}{9} =$ _____

Course:
Instructor:

Name:
Section:

Section 1.2 – Objective 7: Convert Between Percents and Decimals
Part I – Convert a Decimal to a Percent
Video Length – 1:27

Convert a Decimal to a Percent

Multiply the decimal by _____ .

For example, write 0.32 as a percent:

Shortcut: To convert a decimal to a percent, move the decimal _____ places to the _____ and add the percent symbol.

Now write 0.0625 as a percent:

Course: Name:
Instructor: Section:

Section 1.2 – Objective 7: Convert Between Percents and Decimals
Part II – Convert a Percent to a Decimal
Video Length – 3:02

Now we will convert from percents to decimals.

Definition
The word _____ means _____ _____ _____ or _____ out of _____ _____ .

Convert a Decimal to a Percent

                        Multiply the percent by _____ .

For example, write 3% as a decimal:

Shortcut: To convert from a percent to a decimal, drop the percent and move the decimal _____ units to the _____.

So 18% = _____ and 143% = _____ .

Course: Name:
Instructor: Section:

Section 1.3 Video Guide
The Number Systems and the Real Number Line

Objectives:
1. Classify Numbers
2. Plot Points on a Real Number Line
3. Use Inequalities to Order Real Numbers
4. Compute the Absolute Value of a Real Number

Section 1.3 – Introduction
Video Length – 1:56

Definition
A _____ is a collection of objects.

For example, the set of even numbers between 2 and 10, inclusive, can be represented by

$$C = \underline{\hspace{2in}}$$

Definition
When a set has no elements in it, the set is an _____ _____. Empty sets are denoted by _____ or _____ .

Course: Name:
Instructor: Section:

Section 1.3 – Objective 1: Classify Numbers
Video Length – 12:38

Definition
The _____ numbers, or _____ numbers, are the numbers in the set _____.

Definition
The _____ numbers are the numbers in the set _____.

Definition
The _____ are the numbers in the set _____.

Definition
A _____ number is a number that can be expressed as a fraction (or quotient) of two _____ (Note: The integer in the denominator cannot be zero). A rational number can be written in the form $\frac{p}{q}$, where p and q are integers (Note: $q \neq 0$).

Examples of rational numbers:

Additionally, it is important to recognize that $\frac{p}{1} = p$ for any integer. This means that any integer can be written as a _____ number. However, any rational number is not necessarily an integer.

Definition
An _____ number is a number that cannot be written as the quotient of two integers.

We learned in the last section that any fraction can be expressed as a decimal or we can convert decimals to fractions. Likewise, any rational number can also be represented as a decimal. A rational number will have a decimal representation that either _____ (e.g. $\frac{1}{2} = 0.5$) or has a block of numbers that _____ (e.g. $\frac{2}{3} = 0.66666666...$).

With irrational numbers, the decimal representation _____ _____ _____ _____.

Course: Name:
Instructor: Section:

Definition
The set of rational numbers combined with the set of irrational numbers is called the set of

_____ numbers.

The Real Numbers

Rational numbers = _____	**Irrational numbers** = _____
_____	_____
_____	

1. **Example:** List the numbers in the set $\left\{7, -\dfrac{4}{9}, -2, 0, -5.131131113\ldots, 2.\overline{68}, 26.8686\ldots\right\}$ that are

    (a) Counting numbers   (a) **Final answer:** _____

    (b) Whole numbers   (b) **Final answer:** _____

    (c) Integers   (c) **Final answer:** _____

    (d) Rational numbers   (d) **Final answer:** _____

    (e) Irrational numbers   (e) **Final answer:** _____

    (f) Real numbers   (f) **Final answer:** _____

Course: Name:
Instructor: Section:

Section 1.3 – Objective 2: Plot Points on a Real Number Line
Video Length – 5:32

Definition
The real numbers can be represented by points on a line called the _____ _____ _____ .

Definition
The distance between 0 and 1 determines the _____ of the number line. The number associated with a point is called the _____ of the point.

Note: Pay special attention to the subtle difference on how the word "point" and "coordinate" is used.

2. **Example:** On the real number line, label the points with the coordinates 0, 1.75, −2.5, 4.

   Final answer:

Course:
Instructor:

Name:
Section:

Section 1.3 – Objective 3: Use Inequalities to Order Real Numbers
Video Length – 4:25

A real number line allows you to order, or rank, numbers.

3. **Example:** Replace the ? with <, >, or =.

    (a)   3 ? −2   (a) **Final answer:** _____

    (b)   3 ? 5.6   (b) **Final answer:** _____

    (c) -2 ? $-\dfrac{8}{4}$ (c) **Final answer:** _____

    (d) $\dfrac{7}{12}$? $\dfrac{5}{8}$ (d) **Final answer:** _____

Course: Name:
Instructor: Section:

Section 1.3 – Objective 4: Compute the Absolute Value of a Real Number
Video Length – 1:29

Definition
The _____ _____ of a number a, written _____, is the _____ from 0 to a on the real number line.

For example, $|5| =$ _____ :

Also, $|-2| =$ _____ :

Course: Name:
Instructor: Section:

Section 1.4 Video Guide
Adding, Subtracting, Multiplying, and Dividing Integers

Objectives:
1. Add Integers
2. Determine the Additive Inverse of a Number
3. Subtract Integers
4. Multiply Integers
5. Divide Integers

Section 1.4 – Objective 1: Add Integers
Video Length – 8:35

This entire section is dedicated to adding, subtracting, multiplying, and dividing integers. In the next section, we will include rational numbers.

Operations on Signed Numbers
The symbols used in algebra for the operations of addition, subtraction, multiplication, and division are ____, ____, ____, and ____, respectively.

Note: Pay attention to what he says about the division symbol ÷ and mixed numbers.

Operation	Symbols	Words
Addition		
Subtraction		
Multiplication		
Division		

1. **Example:** Add $-5+(-3)$

   Final answer: $-5+(-3)=$ _____

   Now add $-3+2$: ___ :

   And add $5+(-2)=$ ___ :

Course: Name:
Instructor: Section:

Steps to Adding Two Nonzero Integers

To add integers with the same sign (both positive or both negative),
Step 1: Add the absolute value of the two integers.
Step 2: Attach the common sign, either positive or negative.

To add integers with different signs (one positive and one negative),
Step 1: Subtract the smaller absolute value from the larger absolute value.
Step 2: Attach the sign of the integer with the larger absolute value.

2. **Example:** Add $-12+(-42)$

   Final answer: $-12+(-42) =$ _____

3. **Example:** Add $-63+18$

   Final answer: $-63+18 =$ _____

Course:
Instructor:
Name:
Section:

Section 1.4 – Objective 2: Determine the Additive Inverse of a Number
Video Length – 1:16

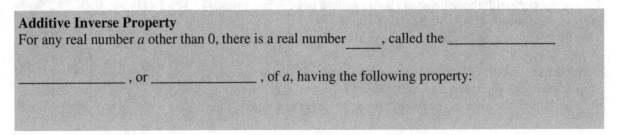

Additive Inverse Property
For any real number *a* other than 0, there is a real number ____ , called the _____

_____ , or _____ , of *a*, having the following property:

What is the additive inverse of 3? ____

What is the additive inverse of –15 ? ____

Course:
Instructor:

Name:
Section:

Section 1.4 – Objective 3: Subtract Integers
Video Length – 3:00

Definition
The _____ $a - b$, read "a _____ b" or "a _____ b," is defined as

In words, to subtract b from a, add the "_____" of b to a.

4. Example: $-16 - 43$

   Final answer: $-16 - 43 =$ _____

5. Example: $17 - (-13)$

   Final answer: $17 - (-13) =$ _____

6. Example: $35 - 81$

   Final answer: $35 - 81 =$ _____

Course: Name:
Instructor: Section:

Section 1.4 – Objective 4: Multiply Integers
Video Length – 5:55

If we multiply two positive numbers together, we get a _____ number. If we multiply two negative numbers together, we get a _____ number. Additionally, if we multiply a positive number and a negative number, we get a _____ number. Why? Instead of just saying, "that's just the rules" or "that's just the way it is", let's provided some justification to those results.

If you're back in an arithmetic class and are trying to explain what $3 \cdot 6$ means, what would you say?

$$3 \cdot 6 =$$

What about $2 \cdot (-9)$?

$$2 \cdot (-9) =$$

But why does the product of two negative numbers give us a positive number? Let's deduce this result by establishing a pattern. For example,

___ · ___ = ___

___ · ___ = ___

___ · ___ = ___

___ · ___ = ___

___ · ___ = ___

Now we have some justification for the rules that we already know.

Rules of Signs for Multiplying Two Integers

1. If we multiply two positive integers, the product is positive.
2. If we multiply one positive integer and one negative integer, the product is negative.
3. If we multiply two negative integers, the product is positive.

7. **Example:** Find the product: $-75 \cdot (-3)$

Final answer: $-75 \cdot (-3) =$ _____

Course:
Instructor:

Name:
Section:

8. **Example:** Find the product: $-18 \cdot 15$

   Final answer: $-75 \cdot (-3) =$ _____

9. **Example:** Find the product: $2 \cdot (-10) \cdot 8$

   When multiplying three or more numbers, multiply from _____ to _____ .

   Final answer: $2 \cdot (-10) \cdot 8 =$ _____

Course: Name:
Instructor: Section:

Section 1.4 – Objective 5: Divide Integers
Video Length – 5:12

Multiplicative Inverse (Reciprocal) Property
For each *nonzero* real number a, there is a real number _____, called the **multiplicative inverse** or **reciprocal** of a, having the following property:

10. **Example:** Find the multiplicative inverse or reciprocal of the following.

 (a) –9 (a) **Final answer:** _____

 (b) $\dfrac{2}{5}$ (b) **Final answer:** _____

So why do we care so much about the reciprocal? Reciprocals will allow us to redefine _____

as _____ .

Definition
If b is a nonzero real number, the _____ $\dfrac{a}{b}$, read as "a _____ by b" or "the _____ of a to b," is defined as

Now that we are able to rewrite division as multiplication, all the rules of signs that apply to multiplication also apply to division.

Rules of Signs for Dividing Two Real Numbers
1. If we divide two positive real numbers, the quotient is _____.

2. If we multiply one positive real number and one negative real number, the quotient is _____.

3. If we multiply two negative real numbers, the quotient is _____.

11. **Example:** Find the quotient: $\dfrac{-75}{15}$
 Note: Pay careful attention to what he says about what remains in the denominator.

 Final answer: $\dfrac{-75}{15} =$ _____

Course: Name:
Instructor: Section:

Section 1.5 Video Guide
Adding, Subtracting, Multiplying, and Dividing Rational Numbers

Objectives:
1. Multiply Rational Numbers in Fractional Form
2. Divide Rational Numbers in Fractional Form
3. Add and Subtract Rational Numbers in Fractional Form
4. Add, Subtract, Multiply, and Divide Rational Numbers in Decimal Form

Section 1.5 – Objective 1: Multiply Rational Numbers in Fractional Form
Video Length – 6:23

Multiplying Fractions

$$\frac{}{} \cdot \frac{}{} = \frac{}{}, \text{ where } b \text{ and } d \neq 0.$$

Note: The approach used in this section will be utilized later on in the course when the numerator and denominator contain variables. So the techniques used here will also apply to future material.

1. **Example:** Find the product: $\left(-\frac{5}{12}\right)\left(-\frac{2}{3}\right)$

   Final answer: $\left(-\frac{5}{12}\right)\left(-\frac{2}{3}\right) = $ _____

2. **Example:** Find the product: $\left(\frac{5}{8}\right)\left(-\frac{20}{7}\right)$

   Final answer: $\left(\frac{5}{8}\right)\left(-\frac{20}{7}\right) = $ _____

Copyright © 2014 Pearson Education, Inc.   23

Course: Name:
Instructor: Section:

Section 1.5 – Objective 2: Divide Rational Numbers in Fractional Form
Video Length – 6:29

Dividing Fractions

$$\frac{}{} \div \frac{}{} = \frac{}{} \cdot \frac{}{} = \frac{}{}, \text{ where } b, c \text{ and } d \neq 0.$$

3. **Example:** Find the quotient: $\left(-\dfrac{4}{9}\right) \div \left(-\dfrac{5}{6}\right)$

   Note: Pay attention to the instructor's "chess" suggestion.

   Final answer: $\left(-\dfrac{4}{9}\right) \div \left(-\dfrac{5}{6}\right) = $ _____

4. **Example:** Find the quotient: $\left(\dfrac{8}{35}\right) \div \left(-\dfrac{1}{10}\right)$

   Final answer: $\left(\dfrac{8}{35}\right) \div \left(-\dfrac{1}{10}\right) = $ _____

Course: Name:
Instructor: Section:

Section 1.5 – Objective 3: Add and Subtract Rational Numbers in Fractional Form Part I
Video Length – 2:43

Adding or Subtracting Fractions with the Same Denominator

$$\frac{a}{c} + \frac{b}{c} = \frac{a+b}{c}, \text{ where } c \neq 0.$$

$$\frac{a}{c} - \frac{b}{c} = \frac{a-b}{c} = \frac{a-b}{c}, \text{ where } c \neq 0.$$

5. **Example:** Find the sum and write in lowest terms: $-\frac{4}{9} + \frac{1}{9}$

Final answer: $-\frac{4}{9} + \frac{1}{9} = $ _____

Course: Name:
Instructor: Section:

Section 1.5 – Objective 3: Add and Subtract Rational Numbers in Fractional Form Part II
Video Length – 7:45

When adding or subtracting two rational number with different denominators, the least common denominator must be found.

Definition
The _____ _____ _____ (_____) is the _____ number that each denominator has as a _____ _____.

We already know how to find the least common denominator of a fraction. The approach to finding the least common denominator of a rational number is identical.

6. **Example:** Find the LCD: $\dfrac{5}{6}$ and $\dfrac{3}{8}$

   Final answer: LCD = _____

7. **Example:** Find the difference: $\dfrac{5}{12} - \dfrac{8}{30}$

   Note: During Step 1, the instructor mentions a method that can be used to find the LCD for fractions with unlike denominators (e.g. using multiples of the larger denominator). However, he indicates that this particular method will not work well for algebra.

Write the steps in words	Show the steps with math
Step 1	
Step 2	
Step 3	
Step 4	

   Final answer: $\dfrac{5}{12} - \dfrac{8}{30} =$ _____

Course: Name:
Instructor: Section:

Section 1.5 – Objective 4: Add, Subtract, Multiply, and Divide Rational Numbers in Decimal Form
Part I
Video Length – 2:47

Adding and Subtracting Decimals
To add or subtract decimals, arrange the numbers in a _____ with the _____ aligned. Then add or subtract the digits in the _____ place values, and place the _____ _____ in the answer directly _____ the decimal point in the problem.

8. **Example:** Find the sum: $718.97 + 496.5$

   Final answer: $718.97 + 496.5 =$ _____

9. **Example:** Find the difference: $8 - 1.623$

   Final answer: $8 - 1.623 =$ _____

Course:
Instructor:

Name:
Section:

Section 1.5 – Objective 4: Add, Subtract, Multiply, and Divide Rational Numbers in Decimal Form
Part II
Video Length – 3:38

Multiplying Decimals
The multiplication of decimals comes from the rules for multiplying numbers written as fractions.

$$\frac{3}{10} \times \frac{4}{100} = \frac{12}{1000} \longrightarrow \underline{} \times \underline{} = \underline{}$$

10. **Example:** Find the product: 0.17×0.4

      Final answer: $0.17 \times 0.4 =$ _____

11. **Example:** Find the product: -2.14×0.03

      Final answer: $-2.14 \times 0.03 =$ _____

Course:
Instructor:
Name:
Section:

Section 1.5 – Objective 4: Add, Subtract, Multiply, and Divide Rational Numbers in Decimal Form
Part III
Video Length – 3:15

The number that is being divided into is called the _____ . The number you are dividing by is called the _____ and the result is called the _____ .

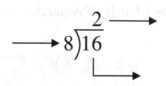

Dividing Decimals
To divide decimals, we want the divisor to be a whole number, so we multiply the dividend and the divisor by a power of 10 that will make the divisor a whole number. Then divide as though we were working with whole numbers. The decimal point in the quotient lies directly above the decimal point in the dividend.

12. **Example:** Find the quotient: $\dfrac{17.68}{13.6}$

Final answer: $\dfrac{17.68}{13.6} =$ _____

Copyright © 2014 Pearson Education, Inc.

Course: Name:
Instructor: Section:

Section 1.6 Video Guide
Properties of Real Numbers

Objectives:
1. Use the Identity Properties of Addition and Multiplication
2. Use the Commutative Properties of Addition and Multiplication
3. Use the Associative Properties of Addition and Multiplication
4. Understand the Multiplication and Division Properties of 0

Section 1.6 – Objective 1: Use the Identity Properties of Addition and Multiplication
Video Length – 4:31

Identity Property of Addition
For any real number a,

$$\_\_ + \_\_ = \_\_ + \_\_ = \_\_$$

That is, the sum of any number and 0 is that number. We call 0 the _____ .

Multiplicative Identity
For any real number a,

$$\_\_ \cdot \_\_ = \_\_ \cdot \_\_ = \_\_$$

That is, the product of any number and 1 is that number. We call 1 the _____ .

Definition

_____ is changing the units of measure from one measure to a different measure.

1. **Example**: Convert 15 feet to inches.

      Final answer: _____

2. **Example**: Convert 3 hours to seconds.

      Final answer: _____

Course:
Instructor:
Name:
Section:

Section 1.6 – Objective 2: Use the Commutative Properties of Addition and Multiplication
Video Length – 10:34

Commutative Property of Addition
If a and b are real numbers, then

$$\underline{} + \underline{} = \underline{} + \underline{}$$

Commutative Property of Multiplication
If a and b are real numbers, then

$$\underline{} \cdot \underline{} = \underline{} \cdot \underline{}$$

3. **Example:** Evaluate the expression: $24 + 7 + (-24)$

   Final answer: $24 + 7 + (-24) = $ _____

By the way, is subtraction commutative?

Is division commutative?

4. **Example:** Evaluate the expression: $\dfrac{2}{3} \cdot 5 \cdot \dfrac{9}{16}$

   Final answer: $\dfrac{2}{3} \cdot 5 \cdot \dfrac{9}{16} = $ _____

5. **Example:** Evaluate the expression: $-15 \cdot 9 \cdot \left(-\dfrac{4}{5}\right)$

   Final answer: $-15 \cdot 9 \cdot \left(-\dfrac{4}{5}\right) = $ _____

Course:
Instructor:
Name:
Section:

Section 1.6 – Objective 3: Use the Associative Properties of Addition and Multiplication
Video Length – 4:47

Associative Property of Addition and Multiplication
If a, b, and c are real numbers, then

$$\underline{}+(\underline{}+\underline{})=(\underline{}+\underline{})+\underline{}=\underline{}+\underline{}+\underline{}$$

$$\underline{}\cdot(\underline{}\cdot\underline{})=(\underline{}\cdot\underline{})\cdot\underline{}=\underline{}\cdot\underline{}\cdot\underline{}$$

6. **Example:** Evaluate the expression: $123+245+(-245)$

Final answer: $123+245+(-245)=$ _____

7. **Example:** Evaluate the expression: $-\dfrac{4}{13}\cdot\dfrac{5}{9}\cdot\dfrac{27}{10}$

Final answer: $-\dfrac{4}{13}\cdot\dfrac{5}{9}\cdot\dfrac{27}{10}=$ _____

"Working smart will save you a boatload of headaches."

Course: Name:
Instructor: Section:

Section 1.6 – Objective 4: Use the Multiplication and Division Properties of Zero
Video Length – 1:54

Multiplication Property of Zero
For any real number a, the product of a and 0 is always 0;

$$\underline{} \cdot \underline{} = \underline{} \cdot \underline{} = \underline{}$$

Division Properties of Zero
For any nonzero real number a,

1. The quotient of 0 and a is 0. That is, $\dfrac{0}{a} = \underline{}$.

2. The quotient of a and 0 is _____ . That is, $\dfrac{a}{0}$ is _____ .

8. **Example:** Find the quotient:

    (a) $\dfrac{23}{0}$ (a) **Final answer:** _____

       Note: The instructor says, "Hey Sullivan! Why the heck is 23 divided by 0 called 'undefined?' I don't understand." Listen carefully to the explanation.

    (b) $\dfrac{0}{17}$ (b) **Final answer:** _____

Course: Name:
Instructor: Section:

Section 1.7 Video Guide
Exponents and Order of Operations

Objectives:
1. Evaluate Exponential Expressions
2. Apply the Rules for Order of Operations

Section 1.7 – Objective 1: Evaluate Exponential Expressions
Video Length – 11:51

Definition
Integer _____ provide a shorthand notation device for representing repeated multiplications of a real number.

$$3 \times 3 \times 3 \times 3 = 3^4 = 81$$

3^4 is read as "_____ to the _____ _____." The 3 is called the _____

and the 4 is called the _____ or _____.

If n is a natural number and a is a real number, then

$$a^n = \underbrace{\phantom{\text{_____}}}$$

The exponent tell the number of times the base is used as a factor.

Note: There is a ton of discussion on part (c) of the following example. Pay very close attention to the follow-up examples and explanations he provides. Also, he does not complete parts (d) and (e). You SHOULD do these on your own.

1. **Example:** Evaluate each expression.

    (a) 2^4 (a) **Final answer:** $2^4 =$ _____

    (b) 5^3 (b) **Final answer:** $5^3 =$ _____

    (c) $(-3)^5$ (c) **Final answer:** $(-3)^5 =$ _____

    (d) $-(12)^2$ (d) **Final answer:** $-(12)^2 =$ _____

    (e) $\left(\dfrac{2}{3}\right)^3$ (d) **Final answer:** $\left(\dfrac{2}{3}\right)^3 =$ _____

Copyright © 2014 Pearson Education, Inc.

Course: Name:
Instructor: Section:

Section 1.7 – Objective 2: Apply the Rules for Order of Operations
Part I – Text Example 5
Video Length – 5:24

We will now look at the order of operations. Suppose you have an expression with multiplication and addition. Which operation do you do first, the multiplication or the addition? _____ . Why? Consider the example $3 \cdot 5 + 8$:

2. **Example:** Evaluate each expression.

   (a) $12 \div 2 - 4 \cdot 2$

   Final answer: $12 \div 2 - 4 \cdot 2 = $ _____

   (b) $2 + 15 \div 5 \cdot 4$
   Note: Remember, multiply/divide from left to right, then add/subtract.

   Final answer: $2 + 15 \div 5 \cdot 4 = $ _____

Course: Name:
Instructor: Section:

Section 1.7 – Objective 2: Apply the Rules for Order of Operations
Part II – Text Examples 6 and 7
Video Length – 7:36

We now understand that we multiply before we add, but what if we wanted to do the addition first? Is there any way that can be accomplished? Consider the expression

$$3 + 2 \cdot 6$$

Now consider

$$(7 + 2) \cdot 4$$

Lastly, consider

$$\left(\frac{3}{4} - \frac{7}{4}\right)\left(\frac{13}{5} + \frac{2}{5}\right)$$

Note: In Example 3, he does not complete part (a). Make sure you do it. Also, pay attention to what he says about the division bar in part (b).

3. **Example:** Evaluate each expression.

    (a) $12 \div 2 - 4 \cdot 2$ 　　　　　　　　　　　　　　(a) **Final answer:** $12 \div 2 - 4 \cdot 2 =$ _____

    (b) $\dfrac{2+6}{16-4 \cdot 3}$ 　　　　　　　　　　　　　(b) **Final answer:** $\dfrac{2+6}{16-4 \cdot 3} =$ _____

4. **Example:** Evaluate the expression: $\dfrac{2 + 3 \div \frac{1}{4}}{-8 \cdot 2 + 9}$

    Final answer: $\dfrac{2 + 3 \div \frac{1}{4}}{-8 \cdot 2 + 9} =$ _____

Course: Name:
Instructor: Section:

Section 1.7 – Objective 2: Apply the Rules for Order of Operations
Part III – Text Example 8
Video Length – 4:53

When you have an expression with multiple grouping symbols, you work _____

_____. Examples of grouping symbols are _____, _____, and

_____.

5. **Example:** Evaluate the expression: $2 \cdot [4 \cdot (3+5) - 7]$

   Final answer: $2 \cdot [4 \cdot (3+5) - 7] =$ _____

6. **Example:** Evaluate the expression: $\left[5 \cdot \left(\frac{3}{4} \cdot (-8) + 2 \right) \right] + \frac{2}{3}$

   Final answer: $\left[5 \cdot \left(\frac{3}{4} \cdot (-8) + 2 \right) \right] + \frac{2}{3} =$ _____

Course: Name:
Instructor: Section:

Section 1.7 – Objective 2: Apply the Rules for Order of Operations
Part IV – Text Examples 9, 10, and 11
Video Length – 14:20

Now, you should ask yourself the following question, "When do we evaluate exponents in the order of operations?"

Consider the expression
$$2 \cdot 4^3$$

The order of operations can be summarized as follows:

1. _____

2. _____

3. _____

4. _____

7. **Example:** Evaluate the expression: $4 + (4^2 - 13)^2 - 3$

   Final answer: $4 + (4^2 - 13)^2 - 3 = $ _____

8. **Example:** Evaluate the expression: $\dfrac{3 \cdot 4^2 - 10}{2(2 - 11)}$

   Final answer: $\dfrac{3 \cdot 4^2 - 10}{2(2 - 11)} = $ _____

Course: Name:
Instructor: Section:

9. **Example:** Evaluate the expression: $\left(\dfrac{6-3^2}{12-2\cdot 4}\right)^2$

Final answer: $\left(\dfrac{6-3^2}{12-2\cdot 4}\right)^2 =$ _____

10. **Example:** Evaluate the expression: $\left(\dfrac{6-(-4)^3}{4^2-2\cdot 3}\right)^2$

    Note: Think "bite-sized chunks".

Final answer: $\left(\dfrac{6-(-4)^3}{4^2-2\cdot 3}\right)^2 =$ _____

Course: Name:
Instructor: Section:

Section 1.8 Video Guide
Simplifying Algebraic Expressions

Objectives:
1. Evaluate Algebraic Expressions
2. Identify Like Terms and Unlike Terms
3. Use the Distributive Property
4. Simplify Algebraic Expressions by Combining Like Terms

Section 1.8 – Objective 1: Evaluate Algebraic Expressions
Video Length – 5:47

Definition
_____ is a branch of mathematics in which symbols represent _____ or

_____ of a set.

Definition
A _____ is a letter used to represent *any* number.

Usually, the variable must come from a predetermined set of possible values. For example, we might say that the variable can be any real number or the variable can only be an integer.

Definition
A _____ is either a _____ number or a letter that represents a fixed number.

Definition
An _____ _____ is any combination of _____ ,

_____ , _____ _____ , and mathematical _____ .

Examples of algebraic expressions:

When you have a number and a variable next to each other without an operation between them, the

operation is understood to be _____ .

Definition
To _____ an _____ _____ , _____ the

numerical value for each variable into the expression and simplify the result.

1. **Example**: Evaluate each expression for the given value.

   (a) $5x - 2$ for $x = 8$ (a) **Final answer:** _____

   (b) $3a^2 + 2a + 4$ for $a = -4$ (b) **Final answer:** _____

Course: Name:
Instructor: Section:

Section 1.8 – Objective 2: Identify Like Terms and Unlike Terms
Video Length – 5:18

Definition
A _____ is a constant or the product or quotient of a constant and one or more variables raised to powers.

Examples of terms:

The term is always going to be the expression separated by the _____ sign.

Algebraic Expression	Terms

Definition
The _____ of a term is the numerical factor of the term.

Definition
Terms that have the same _____ _____(s) with the same _____(s) are called _____ _____.

2. **Example:** Are $4a^2$ and $-7a^2$ like terms? Why or why not? Explain.

3. **Example:** Are $3x^2$ and $-2x^3$ like terms? Why or why not? Explain.

4. **Example:** Are $2ab^2$ and $4a^2b$ like terms? Why or why not? Explain.

5. **Example:** Are 6 and −12 like terms? Why or why not? Explain.

Course:
Instructor:

Name:
Section:

Section 1.8 – Objective 3: Use the Distributive Property
Video Length – 6:02

The Distributive Property
If a, b, and c are real numbers, then

$$\underline{} \cdot (\underline{} + \underline{}) = \underline{} \cdot \underline{} + \underline{} \cdot \underline{}$$

$$(\underline{} + \underline{}) \cdot \underline{} = \underline{} \cdot \underline{} + \underline{} \cdot \underline{}$$

That is, multiply each term inside the parentheses by the factor on the outside.

6. **Example:** Use the Distributive Property to remove the parentheses.

   $7(4+2)$

   Final answer: $7(4+2) = $ _____

7. **Example:** Use the Distributive Property to remove the parentheses.

   $14(23)$

   Final answer: $14(23) = $ _____

8. **Example:** Use the Distributive Property to remove the parentheses.

   $2(a+7)$

   Final answer: $2(a+7) = $ _____

9. **Example:** Use the Distributive Property to remove the parentheses.

   $-\dfrac{1}{4}(12x-16)$

   Final answer: $-\dfrac{1}{4}(12x-16) = $ _____

Course:
Instructor:

Name:
Section:

Section 1.8 – Objective 4: Simplify Algebraic Expressions by Combining Like Terms
Part I – Text Examples 7 and 8
Video Length – 5:21

We will now combine like terms by using the Distributive Property.

10. Example: Simplify each expression by combining like terms.

(a) $4x + 3 + 8x$

Final answer: $4x + 3 + 8x =$ _____

(b) $3a^2 + 2a - 9a^2 + 10a$

Final answer: $3a^2 + 2a - 9a^2 + 10a =$ _____

Course:
Instructor:
Name:
Section:

Section 1.8 – Objective 4: Simplify Algebraic Expressions by Combining Like Terms
Part II – Text Example 9
Video Length – 11:31

11. **Example:** Simplify each expression.

(a) $y+6+8(5-y)$

Final answer: $y+6+8(5-y) = $ _____

(b) $\dfrac{1}{4}\left(\dfrac{2}{3}x-\dfrac{1}{2}\right)+\dfrac{1}{10}\left(\dfrac{5}{2}x-\dfrac{15}{4}\right)$

Note: Be patient. This one takes a while. But you know you love fractions!!!

Final answer: $\dfrac{1}{4}\left(\dfrac{2}{3}x-\dfrac{1}{2}\right)+\dfrac{1}{10}\left(\dfrac{5}{2}x-\dfrac{15}{4}\right) = $ _____

Course: Name:
Instructor: Section:

Section 2.1 Video Guide
Linear Equations: The Addition and Multiplication Properties of Equality

Objectives:
1. Determine Whether a Number Is a Solution of an Equation
2. Use the Addition Property of Equality to Solve Linear Equations
3. Use the Multiplication Property of Equality to Solve Linear Equations

Section 2.1 – Objective 1: Determine Whether a Number Is a Solution of an Equation
Video Length – 5:19

Definition
A _____ _____ in _____ _____ is an equation that can be written in the form ___ + ___ = ___ where a, b, and c are real numbers and $a \neq 0$.

The expressions are called the _____ of the equation.

Equations are _____, statements that can either be _____ or _____. Our goal is to find those values of the variable that make the equation a true statement. Meaning that when we substitute those values of the variable into the equation, the quantity on the left side will be the same quantity that is on the right side.

Definition
The _____ of a linear equation is the value or values of the variable that make the equation a true statement. The set of all solutions of an equation is called the _____ _____. The solution _____ the equation.

1. **Example:** Determine if $x = -1$ is a solution to the equation.

    $$-3(x-3) = -4x + 3 - 5x$$

    Final answer: _____

    If we let $x = 5$ in this equation, the left side is -6 and the right side is -42. Since $-6 \neq -42$, then $x = 5$ is not a solution to the equation. (*Do this yourself!!!*)

Copyright © 2014 Pearson Education, Inc.   45

Course:
Instructor:
Name:
Section:

Section 2.1 – Objective 2: Use the Addition Property of Equality to Solve Linear Equations
Video Length – 11:08

Linear equations are solved by writing a series of steps that result in the equation

$$x = a\ number.$$

One method for solving equations is to write a series of _____ _____.

So how do we form equivalent equations? We utilize different mathematical properties. One such property is called the **Addition Property of Equality**.

Addition Property of Equality
The **Addition Property of Equality** states that for real numbers a, b, and c,

if ___ = ___ , then ___ + ___ = ___ + ___ .

In other words, whatever you add _____

_____

For example, consider the equation

$$6 + y = 11$$

Our goal in solving any linear is to get the variable _____ _____ with the coefficient of _____ .

By the way, when you solve any kind of equation, it is in your best interest to always, always, always,

_____ _____ _____ !!!

2. **Example:** Solve the linear equation $x - 9 = 22$.
   Note: Pay attention to the notation used for the solution set.

   Final answer: _____

3. **Example:** Solve the linear equation $n + \dfrac{4}{3} = \dfrac{6}{5}$.
   Note: A student suggests multiplying by the reciprocal. This idea will not be used here because we are trying to undo addition. However, it will be used later down the road. Stay tuned.

   Final answer: _____

Course: Name:
Instructor: Section:

Section 2.1 – Objective 3: Use the Multiplication Property of Equality to Solve Linear Equations
Video Length – 9:56

We just looked at the addition property of equality to help us solve a linear equation in one variable. Now we will discuss the **Multiplication Property of Equality**.

> **Multiplication Property of Equality**
> The **Multiplication Property of Equality** states that for real numbers a, b, and c, where $c \neq 0$,
>
> if ___ = ___ , then _____ = _____ .

In other words, whatever you multiply_____

_____

For example, consider the equation
$$\frac{1}{7}x = 4.$$

Remember, our goal is to get the x by itself with a coefficient of _____.

4. **Example:** Solve the linear equation $3x = 81$

   Final answer: _____

5. **Example:** Solve the linear equation $-6z = 15$

   Final answer: _____

6. **Example:** Solve the linear equation: $-\dfrac{10}{3} = -\dfrac{5}{6}k$

   Final answer: _____

Course: Name:
Instructor: Section:

Section 2.2 Video Guide
Linear Equations: Using the Properties Together

Objectives:
1. Use the Addition and Multiplication Properties of Equality to Solve Linear Equations
2. Combine Like Terms and Use the Distributive Property to Solve Linear Equations
3. Solve a Linear Equation with the Variable on Both Sides of the Equation
4. Use Linear Equations to Solve Problems

Section 2.2 – Objective 1: Use the Addition and Multiplication Properties of Equality to Solve Linear Equations
Video Length – 6:27

We will now solve linear equations where we need to use both the Addition Property of Equality and the Multiplication Property of Equality.

1. **Example:** Solve the linear equation $3x - 9 = -24$.

Write the steps in words	Show the steps with math
Step 1	
Step 2	
Step 3	

Final answer: _____

2. **Example:** Solve the linear equation $\frac{5}{4}x + 2 = 17$.

Final answer: _____

Course: Name:
Instructor: Section:

Section 2.2 – Objective 2: Combine Like Terms and Use the Distributive Property to Solve Linear Equations

Video Length – 6:19

We will now combine like terms to solve equations.

3. **Example:** Solve the equation $4x + 9 + 2x = -9$.

   Final answer: _____

4. **Example:** Solve the equation $3(4y + 1) - 4 = -13$.

   Final answer: _____

Course: Name:
Instructor: Section:

Section 2.2 – Objective 3: Solve a Linear Equation with the Variable on Both Sides of the Equation
Video Length – 10:47

We will now deal with equations where there are variables on both sides of the equation. When there are variables on both sides of the equation, our goal is to get all the variables on one side and all the constants on the other side.

5. **Example:** Solve the equation $3x - 8 = 2x - 15$.
    Note: Where do you want your variable? On the left or right side? Listen to what he has to say.

    Final answer: _____

6. **Example:** Solve the equation $12 - 2x - 3(x + 2) = 4x + 12 - x$.

Write the steps in words	Show the steps with math
Step 1	
Step 2	
Step 3	
Step 4	
Step 5	

    Final answer: _____

Course: Name:
Instructor: Section:

Summary: Steps for Solving an Equation in One Variable

Step 1: Remove any _____ using the _____ _____ .

Step 2: Combine _____ _____ on each side of the equation.

Step 3: Use the _____ _____ of _____ to get all the

_____ on one side and all _____ on the other side.

Step 4: Use the _____ _____ of _____ to get the

_____ of the variable equal to _____ .

Step 5: _____ the solution to verify that it _____ the _____

equation . *In other words, check your answer to make sure that you did not make any bone

head mistakes!!!*

Course: Name:
Instructor: Section:

Section 2.2 – Objective 4: Use Linear Equations to Solve Problems
Video Length – 2:59

7. **Example:** Jake earns $14 per hour, with hours worked in excess of 40 hours, Jake gets double-time. How many hours of overtime did Jake work if his gross pay was $896?

   To answer the question, solve the equation

   $$40 \cdot 14 + 14(2h) = 896.$$

   Note: Make sure you understand what each term represents. What does the 896 represent? What does the $40 \cdot 14$ represent? What about the $14(2h)$?

   Final answer: _____ .
   Note: Write your final answer as a complete sentence.

Course: Name:
Instructor: Section:

Section 2.3 Video Guide
Solving Linear Equations Involving Fractions and Decimals; Classifying Equations

Objectives:
1. Use the Least Common Denominator to Solve a Linear Equation Containing Fractions
2. Solve Linear Equations Containing Decimals
3. Classify a Linear Equation as an Identity, Conditional, or a Contradiction
4. Use Linear Equations to Solve Problems

Section 2.3 – Objective 1: Use the Least Common Denominator to Solve a Linear Equation Containing Fractions
Video Length – 10:40

1. **Example:** Solve the equation $\frac{2}{5} + v = \frac{1}{2} - \frac{3}{10}$.

Write the steps in words	Show the steps with math
Step 1	
Step 2	
Step 3	
Step 4	
Step 5 *Note: Remember, you should ALWAYS check your answer. Do it here before you move on.*	

Final answer: _____

Course:
Instructor:

Name:
Section:

2. **Example:** Solve the equation $\dfrac{3n+5}{2} = 4 + \dfrac{4n+2}{4}$

Final answer: _____

Course:　　　　　　　　　　　　　　　　　　　　　　Name:
Instructor:　　　　　　　　　　　　　　　　　　　　　Section:

Section 2.3 – Objective 2: Solve a Linear Equation Containing Decimals
Video Length – 13:22

3. **Example:** Solve the equation $0.4z - 6 = 1.2$.

　　Final answer: _____

4. **Example:** Solve the equation $0.04x + 0.06x(10,000 - x) = 480$.

　　Final answer: _____

Course: Name:
Instructor: Section:

Section 2.3 – Objective 3: Classify a Linear Equation as an Identity, Conditional, or a Contradiction

Video Length – 11:49

Definition
A _____ _____ is an equation that is true for some values of the variable and false for other values of the variable.

A _____ is an equation that is false for every replacement value of the variable.

5. **Example:** Solve the equation $4x - 8 = 4x - 1$.

   Final answer: _____

Definition
An _____ is an equation that is satisfied for all values of the variable for which both sides of the equation are defined.

6. **Example:** Solve the equation $5x + 3 = 2x + 3(x + 1)$.

   Final answer: _____

7. **Example:** Solve and classify the equation $3(2 - 4n) + 1 = -\frac{1}{2}(24n - 6)$.

   Final answer: _____

If you are asked to *classify* the equation, then you will need to state that it is either a contradiction, a conditional equation, or an identity. If you are asked to *solve* the equation, then you will need to state the actual solution set.

Course: Name:
Instructor: Section:

Section 2.4 Video Guide
Evaluating Formulas and Solving Formulas for a Variable

Objectives:
1. Evaluate a Formula
2. Solve a Formula for a Variable

Section 2.4 – Objective 1: Evaluate a Formula
Part I – Text Example 3
Video Length – 6:34

Definition
A mathematical _____ is an equation that describes how two or more variables are related.

Definition
_____ is money paid for the use of money.

Definition
The total amount borrowed is called the _____.

Definition
The _____ ____ _____ , expressed as a percent, is the amount charged for the use of the principal for a given period of time, usually on a yearly basis.

Simple Interest Formula
If an amount of money, P, called the principal is invested for a period of t years at an annual interest rate r, expressed as a decimal, the interest I earned is

$$I = \underline{} \cdot \underline{} \cdot \underline{}$$

This interest earned is called **simple interest**.

1. **Example:** Jordan received a bonus check for $1000. He invested it in a mutual fund that earned 6.75% simple interest. Find the amount of interest Jordan will earn in one month.

Final answer: _____

Course: Name:
Instructor: Section:

Section 2.4 – Objective 1: Evaluate a Formula
Part II – Text Examples 4, 5, and 6
Video Length – 14:14

Definition

The _____ is the sum of the lengths of all the sides of a figure.

The _____ is the amount of space enclosed by a two-dimensional figure measured in square units.

The _____ _____ of a solid is the sum of the areas of the surfaces of a three-dimensional figure.

The _____ is the amount of space occupied by a figure measured in cubic units.

The _____ r of a circle is the line segment that extends from the center of the circle to any point on the circle.

The _____ of a circle is any line segment that extends from one point on the circle through the center to a second point on the circle. The diameter is two times the length of the radius, ___ = ___ .

In circles, we use the term _____ to mean the perimeter.

Plane Figure Formulas

Figure	Formulas
Square	**Area:** $A =$ **Perimeter:** $P =$
Rectangle	**Area:** $A =$ **Perimeter:** $P =$
Triangle	**Area:** $A =$ **Perimeter:** $P =$
Trapezoid	**Area:** $A =$ **Perimeter:** $P =$

(*continued*)

Course: Name:
Instructor: Section:

Plane Figure Formulas (*continued*)

Figure	Formulas
Parallelogram	**Area:** $A =$ **Perimeter:** $P =$
Circle	**Area:** $A =$ **Perimeter:** $P =$

Solid Formulas

Figure	Formulas
Cube	**Volume:** $V =$ **Surface Area:** $S =$
Rectangular Box	**Volume:** $V =$ **Surface Area:** $S =$
Sphere	**Volume:** $V =$ **Surface Area:** $S =$
Right Circular Cylinder	**Volume:** $V =$ **Surface Area:** $S =$
Cone	**Volume:** $V =$

2. **Example:** If the area of a triangle is 66 square inches, and the base is 8 inches, find the height of the triangle.

Final answer: _____

Course: Name:
Instructor: Section:

3. **Example:** You have the opportunity to buy an 18" pizza at a cost of $16.99 or you can buy a 12" pizza at a cost of $9.99. Which is the better buy?

Final answer: _____

Course:
Instructor:
Name:
Section:

Section 2.4 – Objective 2: Solve a Formula for a Variable
Video Length – 10:46

"Solve for a variable" means to get the variable by itself with a coefficient of 1 on one side of the equation and all other variables and constants, if any, on the other side.

4. **Example:** The formula for the area of a trapezoid is $A = \dfrac{h(a+b)}{2}$. Solve the formula for b.

Final answer: _____

5. **Example:** Solve for y: $4x + 3y = 12$.

   Note: Solving for y in this type of equation will show up in future sections. Also, pay attention to the special way he writes the final answer in $y = mx + b$ form, where m and b are real numbers. Again, this will show up in future sections.

Final answer: _____

Course: Name:
Instructor: Section:

Section 2.5 Video Guide
Problem Solving: Direct Translation

Objectives:
1. Translate English Phrases to Algebraic Expressions
2. Translate English Sentences to Equations
3. Build Models for Solving Direct Translation Problems

Section 2.5 – Objective 1: Translate English Phrases to Algebraic Expressions
Video Length – 6:33

Before we translate English sentences to equations, we will first translate English phrases to algebraic expressions. Remember, the difference between an algebraic equation and an algebraic expressions is that an algebraic expression does not contain an equal sign.

Math Symbols and the Words They Represent			
Add (+)	**Subtract (−)**	**Multiply (·)**	**Divide (/)**

1. **Example:** Express each English phrase using mathematical symbols.
   Note: Be careful with part (d).

   (a) The sum of 3 and 8 (a) **Final answer**: _____

   (b) 7 less than 12 (b) **Final answer**: _____

   (c) The quotient of 24 and 6 (c) **Final answer**: _____

   (d) Four times the sum of a number n and 3 (d) **Final answer**: _____

Course: Name:
Instructor: Section:

Section 2.5 – Objective 2: Translate English Sentences to Equations
Part I
Video Length – 3:11

Now we will translate English sentences to algebraic equations.

2. **Example:** Translate each of the following into a mathematical statement.

   (a) Twelve more than a number is 25.

   Final answer: _____

   (b) One-third of the sum of a number and four yields 6.

   Final answer: _____

Course: Name:
Instructor: Section:

Section 2.5 – Objective 2 - Translate English Sentences to Equations
Part II – Introduction to Problem Solving
Video Length – 7:52

Definition
_____ _____ is the ability to use information, tools, and our own skills to achieve a goal.

Definition
The process of taking a verbal description of the problem and developing it into an equation that can be used to solve the problem is _____ _____ .

Definition
The equation that is developed is the _____ _____ .

Five Categories of Problems

1. _____ _____ – problems that must be translated from English into mathematics using key words in the verbal description

2. _____ – problems where _____ or _____ quantities are

3. combined in some fashion

4. _____ – problems where the unknown quantities are related through _____ formulas

5. _____ _____ – problems where an object travels at a _____ speed

6. _____ _____ – problems where _____ or _____ entities join forces to _____ ____ _____

Course: Name:
Instructor: Section:

Solving Problems with Mathematical Models

Step 1: Identify What You Are Looking For
Read the problem carefully. Identify the type of problem and the information we wish to learn. Typically the last sentence in the problem indicates what it is we wish to solve for.

Step 2: Give Names to the Unknowns
Assign variables to the unknown quantities. Choose a variable that is representative of the unknown quantity it represents. For example, use t for time.

Step 3: Translate into the Language of Mathematics
Determine if each sentence can be translated into a mathematical statement. If necessary, combine the statements into an equation that can be solved.

Step 4: Solve the Equation(s) Found in Step 3
Solve the equation for the variable and then answer the question posed by the original problem.

Step 5: Check the Reasonableness of Your Answer
Check your answer to be sure that it makes sense. If it does not, go back and try again.

Step 6: Answer the Question
Write your answer in a complete sentence.

Course:
Instructor:

Name:
Section:

Section 2.5 – Objective 3: Build Models for Solving Direct Translation Problems
Part I – Text Example 6
Video Length – 7:14

3. **Example:** The sum of three consecutive integers is 174. Find the integers.

Write the steps in words	Show the steps with math
Step 1	
Step 2	
Step 3	
Step 4	
Step 5 *Note: He does this step mentally. You should do that AND write down the work as well.*	
Step 6	

Course:
Instructor:

Name:
Section:

Section 2.5 – Objective 3: Build Models for Solving Direct Translation Problems
Part II – Text Example 8
Video Length – 6:36

4. **Example:** A total of $17,500 is to be invested, some in CD's and $4000 less than that in bonds. How much is to be invested in each type of investment?

Write the steps in words	Show the steps with math
Step 1	
Step 2	
Step 3	
Step 4	
Step 5 *Note: He does this step mentally. You should do that AND write down the work as well.*	
Step 6	

Course:
Instructor:

Name:
Section:

Section 2.5 – Objective 3: Build Models for Solving Direct Translation Problems
Part III – Text Example 9
Video Length – 7:46

5. **Example:** EZ Rider Truck Rental charges $50 per day plus $0.25 per mile to rent a small truck. Trucks-R-Us charges $35 per day plus $0.30 per mile for the same model truck. For how many miles will the daily costs of renting the two trucks be the same?

Write the steps in words	Show the steps with math
Step 1	
Step 2	
Step 3	
Step 4	
Step 5 *Note: He does this step mentally. You should do that AND write down the work as well.*	
Step 6	

Course: Name:
Instructor: Section:

Section 2.6 Video Guide
Problem Solving: Problems Involving Percent

Objectives:
1. Solve Problems Involving Percent
2. Solve Business Problems That Involve Percent

Section 2.6 – Objective 1: Solve Problems Involving Percent
Video Length – 3:08

We know that percent means "per one-hundred" or "divided by 100". So if you ever see a percent in a problem, the first thing you will have to do is to convert that percent to a decimal. Basically that means taking the number and dividing it by 100 and dropping the percent sign. Of course we know that dividing by 100 essentially means moving the decimal two places to the left.

We now have a new word that means multiplication. The word "_____" in mathematics means "_____".

1. **Example:** A number is 9% of 65. Find the number.

   Final answer: _____

2. **Example:** 36 is 6% of what number?

   Final answer: _____

Course: Name:
Instructor: Section:

Section 2.6 – Objective 2: Solve Business Problems That Involve Percent
Video Length – 12:09

One type of percent problem involves discounts or mark-ups that businesses use in determining their prices.

_____ _____ − _____ = _____ _____

_____ _____ + _____ = _____ _____

Sales tax is also another example of a problem that involves a percent. We will now look at an example involving sales tax.

3. **Example:** Julie bought a leather sofa that was on sale for 35% off the original price. If she paid $780, what was the original price of the sofa?

Write the steps in words	Show the steps with math
Step 1	
Step 2	
Step 3	
Step 4 Note: He shows the work for this step, however he does not designate it as "④".	
Step 5 Note: He does this step mentally. You should do that AND write down the work as well.	
Step 6	

Course: Name:
Instructor: Section:

4. Example: You just purchased a new jacket (*congratulations!!!*). The price of the jacket, including sales tax of 7%, was $48.15. How much did the jacket cost excluding sales tax?

Write the steps in words	*Show the steps with math*
Step 1	
Step 2	
Step 3	
Step 4	
Step 5 *Note: He does this step mentally. You should do that AND write down the work as well.*	
Step 6	

Course: Name:
Instructor: Section:

Section 2.7 Video Guide
Problem Solving: Geometry and Uniform Motion

Objectives:
1. Set Up and Solve Complementary and Supplementary Angle Problems
2. Set Up and Solve Angles of Triangle Problems
3. Use Geometry Formulas to Solve Problems
4. Set up and Solve Uniform Motion Problems

Section 2.7 – Objective 1: Set Up and Solve Complementary and Supplementary Angle Problems
Video Length – 4:34

Definition
Two angles whose sum is _____ are called _____ _____. Each angle is called the _____ of each other.

Two angles whose sum is _____ are called _____ _____. Each angle is called the _____ of each other.

1. **Example:** Angle A and angle B are complementary angles, and angle A is $21°$ more than twice angle B. Find the measure of both angles.

Write the steps in words	Show the steps with math
Step 1	
Step 2	

(continued)

Course: Name:
Instructor: Section:

Example continued:

Step 3	
Step 4	
Step 5	
Step 6	

Course: Name:
Instructor: Section:

Section 2.7 – Objective 2: Set Up and Solve "Angles of a Triangle" Problems
Video Length – 6:58

The sum of the measures of the interior angles of a triangle is $180°$.

2. **Example:** The measure of the smallest angle of a triangle is 40 degrees less than the measure of the middle angle and the measure of the largest angle is 25 degrees larger than the measure of the middle angle. Find the measure of each angle of the triangle.

Write the steps in words	Show the steps with math
Step 1	
Step 2	
Step 3	
Step 4	
Step 5	
Step 6	

Course: Name:
Instructor: Section:

Section 2.7 – Objective 3: Use Geometry Formulas to Solve Problems
Video Length – 4:08

3. **Example:** The perimeter of a rectangular swimming pool is 64 feet. The length of the pool is 8 feet more than the width. Find the dimensions of the pool.

Write the steps in words	Show the steps with math
Step 1	
Step 2	
Step 3	
Step 4	
Step 5	
Step 6	

Course: Name:
Instructor: Section:

Section 2.7 – Objective 4: Set Up and Solve Uniform Motion Problems
Video Length – 7:13

Definition
Objects that move at a constant velocity are said to be in _____ _____.

When the average velocity of an object is known, it can be interpreted as its constant velocity.

Uniform Motion Formula
If an object moves at an average speed r, the distance d covered in time t is given by the formula

$$\underline{} = \underline{} \cdot \underline{}.$$

4. **Example:** Tanya, who is a long-distance runner, runs at an average speed of 8 miles per hour. Two hours after Tanya leaves your house, you leave in a car and follow the same route. If your average speed is 40 miles per hour, how long will it be before you catch up with Tanya?

Write the steps in words	Show the steps with math
Step 1	
Step 2	
Step 3	
Step 4	
Step 5	
Step 6	

Course: Name:
Instructor: Section:

Section 2.8 Video Guide
Solving Linear Inequalities in One Variable

Objectives:
1. Graph Inequalities on a Real Number Line
2. Use Interval Notation
3. Solve Linear Inequalities Using Properties of Inequalities
4. Model Inequality Problems

Section 2.8 – Objective 1: Graph Inequalities on a Real Number Line
Video Length – 9:23

Definition
A _____ _____ in _____ _____ is an inequality that can be written in the form

_____ or _____ or _____ or _____

where a, b, and c are real numbers and $a \neq 0$.

Examples:

Definition
Inequalities that contain one inequality symbol are called _____ _____.

Note: Our focus will be on simple inequalities. Inequalities with more than one inequality symbol, or **compound inequalities**, will be introduced in a later chapter.

For example, the simple inequality

$x > 4$ means "_____"

Ultimately, we are going to find solutions to linear inequalities. Therefore we will need a way to represent our solutions. There are a number of ways to represent solutions to linear inequalities. One such method is present below.

Definition
_____-_____ _____ is used to express the inequality in written form.

$$\{ \ x \ | \ x > 4 \ \}$$

_____   _____

Course: Name:
Instructor: Section:

Definition
Representing an inequality on a number line is called graphing the inequality, and the picture is called the _____ of the _____.

$$x > 4$$

⟵─────────────────────⟶

1. **Example:** Graph each inequality on a real number line.
   Note: Pay attention to how parentheses and brackets are used when graphing inequalities.

   (a) $x > 3$

   Final answer: ─────────────⟶

   (b) $x \leq -2$

   Final answer: ─────────────⟶

78 Copyright © 2014 Pearson Education, Inc.

Course: Name:
Instructor: Section:

Section 2.8 – Objective 2: Use Interval Notation
Video Length –12:24

We can represent inequalities using set-builder notation and by graphing on a real number line. A third approach to represent inequalities is called **interval notation**.

$$x > 4 \longrightarrow$$

Set-Builder Notation	Interval Notation	Graph

2. **Example:** Graph each inequality on a number line, and write each inequality in set-builder notation and interval notation.

   (a) $x > 5$

   Set-builder: _____

   Interval notation: _____

   Graph: _____

   (b) $x \leq -1$

   Set-builder: _____

   Interval notation: _____

   Graph: _____

Just for fun, represent the following graph in set-builder and interval notation.

$\longrightarrow$

Course: Name:
Instructor: Section:

Section 2.8 – Objective 3: Solve Linear Inequalities Using Properties of Inequalities
Part I – Text Examples 3, 4, 5, and 6
Video Length –10:18

Now we will solve inequalities. To solve an inequality means to find all values of the variable that make the inequality a true statement.

Addition Property of Inequality
For real numbers a, b, and c

If ___ < ___ , then ___ + ___ < ___ + ___

If ___ > ___ , then ___ + ___ > ___ + ___

3. **Example:** Solve the linear inequality and state the solution set using set-builder notation and interval notation. Graph the solution set.

    $x + 24 > 19$

    Set-builder: _____

    Interval notation: _____

    Graph: ⎯⎯⎯⎯⎯⎯➤

Multiplication Properties of Inequality
Let a, b, and c be real numbers.

If ___ < ___ , and if ___ > ___ , then _____ < _____

If ___ > ___ , and if ___ > ___ , then _____ > _____

If ___ < ___ , and if ___ < ___ , then _____ > _____

If ___ > ___ , and if ___ < ___ , then _____ < _____

In other words, if we multiply both sides of an inequality by a positive number, the direction of the inequality does not change. If we multiply both sides of an inequality by a negative number, the direction of the inequality changes.

4. **Example:** Solve the linear inequality and state the solution set using set-builder notation and interval notation. Graph the solution set.

    $-2y < 16$

    Set-builder: _____

    Interval notation: _____

    Graph: ⎯⎯⎯⎯⎯⎯➤

Course:
Instructor:

Name:
Section:

Section 2.8 – Objective 3: Solve Linear Inequalities Using Properties of Inequalities
Part II – Text Examples 7, 8, 9, and 10
Video Length –14:24

5. **Example:** Solve the linear inequality and state the solution set using set-builder notation and interval notation. Graph the solution set.

$$5(2x-1)-6x \leq 11x-19$$

Note: He completes Step 3 in two ways.

Write the steps in words	Show the steps with math
Step 1	
Step 2	
Step 3	
Step 4	

Set-builder: _____

Interval notation: _____

Graph: ⎯⎯⎯⎯⎯⎯⎯⎯⎯➤

Course:
Instructor:

Name:
Section:

Note: This example is a good one!

6. **Example:** Solve the linear inequality and state the solution set using set-builder notation and interval notation. Graph the solution set.

$$2(3x-2)+7>10x-(4x+5)$$

Set-builder: _____

Interval notation: _____

Graph: ───────────▶

7. **Example:** Solve the linear inequality and state the solution set using set-builder notation and interval notation. Graph the solution set.

$$6\left(1-\frac{1}{3}x\right)+5x \leq 3x-4$$

Set-builder: _____

Interval notation: _____

Graph: ───────────▶

Course: Name:
Instructor: Section:

Section 2.8 – Objective 4: Model Inequality Problems
Video Length –7:34

We will now look at problems that will lead to linear inequalities. However, we will first consider certain key words or phrases and how they translate into inequality symbols.

Word or Phrase	Inequality Symbol

Course:
Instructor:

Name:
Section:

8. **Example:** Federation Express will not deliver a package if its height plus girth (circumference around the widest part) is more than 130 inches. If you are preparing a package that is 33 inches wide and 8 inches long, how high is the package permitted to be?

Write the steps in words	*Show the steps with math*
Step 1	
Step 2	
Step 3	
Step 4	
Step 5 Note: He does this step mentally. You should do that AND write down the work as well.	
Step 6	

Course: Name:
Instructor: Section:

Section 3.1 Video Guide
The Rectangular Coordinate System and Equations in Two Variables

Objectives:
1. Plot Points in the Rectangular Coordinate System
2. Determine If an Ordered Pair Satisfies an Equation
3. Create a Table of Values That Satisfy an Equation

Section 3.1 – Objective 1: Plot Points in the Rectangular Coordinate System
Video Length – 12:35

The Rectangular Coordinate System

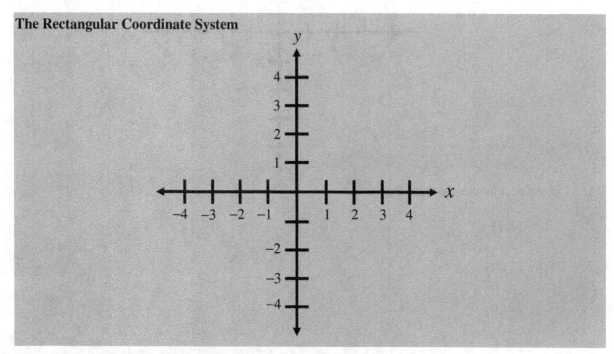

The rectangular coordinate system is also known as the **Cartesian Plane** or the ***xy*-plane**.

Definition
Any point P can be represented by using an _____ _____ (x, y) of real numbers.

Definition
The coordinate system can be divided into four separate regions, or _____ .

Course: Name:
Instructor: Section:

The coordinates of each point tell us how many units from the origin to travel.

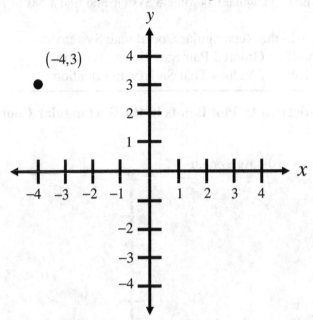

1. **Example:** Plot the points on the xy-plane.

   (a) $(4, 2)$

   (b) $(3, -3)$

   (c) $(-1, 0)$

2. **Example:** Identify the coordinates of each point

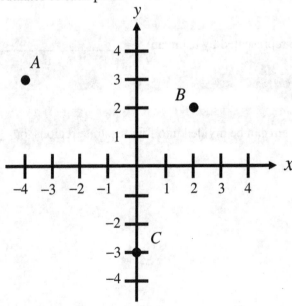

Course: Name:
Instructor: Section:

Section 3.1 – Objective 2: Determine If an Ordered Pair Satisfies an Equation
Video Length – 5:42

Remember back in Chapter 2, we solved an equation in one variable. Then we learned that there are three different categories of linear equations – conditional equations, identities, and contradictions. Now, we will look at equations in two variables.

Definition
An _____ in _____ _____ , x and y, is a statement in which the algebraic expressions involving x and y are equal. The expressions are called _____ of the equation.
Examples:

Definition
Any values of the variables that make the equation a true statement are said to _____ the equation.

Consider the equation
$$x + y = 15$$
The ordered pair $(5,10)$ *satisfies* the equation. Is this the only value of x and y that satisfy the equation?

3. **Example:** Determine if the following ordered pairs satisfy the equation $2x + y = 5$.

    (a) $(2,1)$ (b) $(3,-4)$

    (a) **Final answer:** _____ (b) **Final answer:** _____

Course: Name:
Instructor: Section:

Section 3.1 – Objective 3: Create a Table of Values That Satisfy an Equation
Video Length – 7:48

In the previous section, we learned how to show whether or not a specific ordered pair satisfies an equation. In this section, we will create a table of values that satisfy an equation. In order to do this, we pick values of one of the variables, then use the equation to find the corresponding value of the other variable.

4. **Example:** Use the equation $y = 6x + 5$ to complete the table and list the ordered pairs that satisfy the equation.

x	y	(x, y)
-2		
0		
2		

5. **Example:** Find ordered pairs that satisfy the equation $3x - 2y = -6$.

   Note: A student suggests an alternative method that can be used to complete this problem. You should try it after watching this video.

Ultimately, this procedure of creating a table of values that satisfy an equation will be used to graph the equation. Stay tuned.

Course: Name:
Instructor: Section:

Section 3.2 Video Guide
Graphing Equations in Two Variables

Objectives:
1. Graph a Line by Plotting Points
2. Graph a Line Using Intercepts
3. Graph Vertical and Horizontal Lines

Section 3.2 – Objective 1: Graph a Line by Plotting Points
Part I – Text Example 1
Video Length – 6:28

In the last section, we learned how to find values of x and y that satisfy an equation in two variables. What we are going to do now is take that information and learn how to obtain the graph of an equation in two variables.

Definition
The _____ of an _____ in _____ _____ x and y is the set of all ordered pairs (x, y) in the xy-plane that satisfy the equation.

In other words, the graph of an equation in two variables represents the _____ _____ of the equation.

So how do we obtain the graph of an equation? One method for graphing an equation is the **point-plotting method**.

1. **Example**: Graph the equation $y = x + 3$ by plotting points.

Write the steps in words	Show the steps with math
Step 1	
Step 2	
Step 3	

Course: Name:
Instructor: Section:

Graphing an Equation Using the Point-Plotting Method

Step 1: Find _____ ordered pairs that _____ the equation.

Step 2: _____ the points found in Step 1 in a rectangular coordinate system.

Step 3: _____ the points in a _____ _____ or _____.

Course: Name:
Instructor: Section:

Section 3.2 – Objective 1: Graph a Line by Plotting Points
Part II – Text Example 2
Video Length – 3:01

Definition
The _____ _____ in _____ _____ is an equation of the form

where A, B, and C are real numbers. A and B cannot both be 0.

Definition
When a linear equation is written in the form ___ + ___ = ___ , we can say that the linear equation is

in _____ _____ .

Examples:

Definition
The graph of a linear equation is a _____ .

Course: Name:
Instructor: Section:

Section 3.2 – Objective 1: Graph a Line by Plotting Points
Part III – Text Example 3
Video Length – 13:53

2. **Example:** Graph the equation $2x + 3y = 8$ by plotting points.
   Note: Plot nice points (e.g. integer values for BOTH x and y).

Course: Name:
Instructor: Section:

Section 3.2 – Objective 2: Graph a Line Using Intercepts
Part I – Text Example 5
Video Length – 3:46

Definition

The _____ are the points, if any, where a graph crosses or touches the coordinate axes.

The *x*-coordinate of a point at which the graph crosses or touches the *x*-axis is an

___ - _____, and the *y*-coordinate of a point at which the graph crosses or touches the

y-axis is a ___ - _____ .

Course: Name:
Instructor: Section:

Section 3.2 – Objective 2: Graph a Line Using Intercepts
Part II – Text Examples 6, 7, and 8
Video Length – 11:02

Procedure for Finding Intercepts

1. To find the *x*-intercept(s), if any, of the graph of an equation, let ___ = ___ in the equation and solve for _____ .

2. To find the *y*-intercept(s), if any, of the graph of an equation, let ___ = ___ in the equation and solve for _____ .

3. **Example:** Graph the equation $2x - 3y = 6$ by finding its intercepts.

Write the steps in words	Show the steps with math
Step 1	
Step 2	
Step 3	
Step 4 *Plot the points found in Steps 1-3, and draw the line.*	

Course: Name:
Instructor: Section:

4. **Example:** Graph the equation $3x+4y=0$ by finding its intercepts.

Course: Name:
Instructor: Section:

Section 3.3 Video Guide
Slope

Objectives:
1. Find the Slope of a Line Given Two Points
2. Find the Slope of Vertical and Horizontal Lines
3. Graph a Line Using Its Slope and a Point on the Line
4. Work with Applications of Slope

Section 3.3 – Objective 1: Find the Slope of a Line Given Two Points
Video Length – 15:22

Definition
The _____ of a line, denoted by the letter _____, is the ratio of the _____ to the _____. That is,

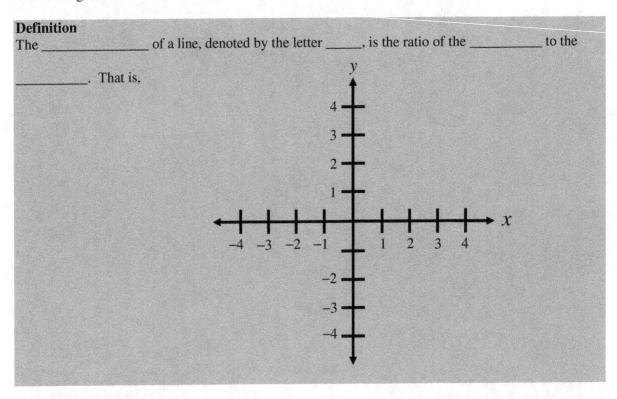

Definition
If $x_1 \neq x_2$, the _____ _____, of the line containing points (x_1, y_1) and (x_2, y_2) is defined by the formula

We can also write the slope m of a line as

Course: Name:
Instructor: Section:

1. **Example:** Find the slope of the line that passes through $(1,2)$ and $(3,6)$.

   Final answer: _____

2. **Example:** Find the slope of the line joining $(3,2)$ and $(-1,10)$.

   Final answer: _____

Course:
Instructor:

Name:
Section:

Section 3.3 – Objective 2: Find the Slope of Vertical and Horizontal Lines
Video Length – 4:50

3. **Example:** Find the slope of the line that passes through $(3,4)$ and $(3,-1)$.

Final answer: _____

In a _____ line, the slope is _____ .

4. **Example:** Find the slope of the line that passes through $(-3,3)$ and $(2,3)$.

Final answer: _____

The slope of a _____ line is _____ .

98   Copyright © 2014 Pearson Education, Inc.

Slope of a Straight Line

Positive Slope
Line _____ from _____ to _____

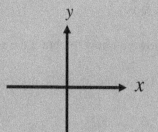

Negative Slope
Line _____ from _____ to _____

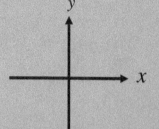

Zero Slope
_____ Line

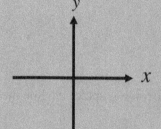

Undefined Slope
_____ Line

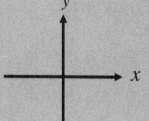

Course: Name:
Instructor: Section:

Section 3.3 – Objective 3: Graph a Line Using Its Slope and a Point on the Line
Video Length – 6:05

We will now show you how to obtain the graph of a line given its slope and a point on the line.

5. **Example:** Draw a graph of the line that contains the point $(-1,3)$ and has slope 2.

6. **Example:** Draw a graph of the line that contains the point $(-2,4)$ and has a slope of $-\dfrac{1}{3}$.

100 Copyright © 2014 Pearson Education, Inc.

Course:
Instructor:
Name:
Section:

Section 3.3 – Objective 4: Work with Applications of Slope
Video Length – 3:49

7. **Example:** Penny decided to take a hike up a mountain trail. The trail has a vertical rise of 90 feet for every 250 feet of horizontal change. In percent, what is the grade of the trail?

Final answer: _____

Course: Name:
Instructor: Section:

Section 3.4 Video Guide
Slope-Intercept Form of a Line

Objectives:
1. Use the Slope-Intercept Form to Identify the Slope and *y*-Intercept of a Line
2. Graph a Line Whose Equation Is in Slope-Intercept Form
3. Graph a Line Whose Equation Is in the Form $Ax + By = C$
4. Find the Equation of a Line Given Its Slope and *y*-intercept
5. Work with Linear Models in Slope-Intercept Form

Section 3.4 – Objective 1: Use the Slope-Intercept Form to Identify the Slope and *y*-Intercept of a Line
Video Length – 10:11

We have defined an equation of a line as $Ax + By = C$, where *A* and *B* cannot both be zero. We also looked at two different techniques for obtaining the graph of the line – point plotting and using intercepts. Additionally, we also looked at how we could obtain the graph of a line if we know a point on the line and the slope of the line. Now, we will develop a technique that will allow us to graph the equation of a line knowing the role slope plays in helping us get that graph.

Consider the equation $3x + y = 7$. If you were asked to graph this line, one technique that you might use is to pick values of *x* and then use the equation to find the corresponding values of *y*.

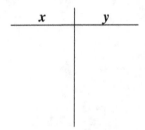

Slope-Intercept Form of an Equation of a Line
An equation of a line with slope _____ and *y*-intercept _____ is

1. **Example:** Find the slope and *y*-intercept of the line whose equation is $x + 2y = 8$.

    Final answer: _____

102   Copyright © 2014 Pearson Education, Inc.

Course:
Instructor:

Name:
Section:

Section 3.4 – Objective 2: Graph a Line Whose Equation Is in Slope-Intercept Form
Video Length – 4:05

2. **Example:** Graph the line $y = \dfrac{3}{2}x - 4$ using the slope and y-intercept.

Write the steps in words	Show the steps with math
Step 1	
Step 2	

Course:
Instructor:

Name:
Section:

Section 3.4 – Objective 3: Graph a Line Whose Equation Is in the Form $Ax + By = C$
Video Length – 5:11

3. **Example:** Graph the line $4x + 3y = 12$ using the slope and y-intercept.

Write the steps in words	Show the steps with math
Step 1	
Step 2	
Step 3 *Note: In regards to the slope, he initially places the negative in the numerator (with the positive in the denominator). Then, as an aside, he places the negative in the denominator (with a positive numerator). Did this change the graph of the line?*	

Course: Name:
Instructor: Section:

Section 3.4 – Objective 4: Find the Equation of a Line Given Its Slope and *y*-intercept
Video Length – 2:17

Up to this point, we were asked to graph a line given its equation. We will now "flip-flop" the process and find the equation of a line given some information about the line (e.g. the slope and *y*-intercept).

4. **Example:** Find the equation of a line with slope $\frac{2}{5}$ and *y*-intercept -2.

Final answer: _____

Now graph the line for fun!

Course:
Instructor:
Name:
Section:

Section 3.4 – Objective 5: Work with Linear Models in Slope-Intercept Form
Video Length – 6:48

5. **Example:** AT&T has a land-line phone plan that charges $0.04 per minute plus an $11 monthly fee.

   (a) Write a linear equation that relates the monthly fee, y, to the minutes used, x.

   Final answer: _____

   (b) What is the fee if you use the phone 100 minutes?

   Final answer: _____

   (c) Graph.

Course: Name:
Instructor: Section:

Section 3.5 Video Guide
Point-Slope Form of a Line

Objectives:
1. Find the Equation of a Line Given a Point and a Slope
2. Find the Equation of a Line Given Two Points
3. Build Linear Models Using the Point-Slope Form of a Line

Section 3.5 – Objective 1: Find the Equation of a Line Given a Point and a Slope
Video Length – 6:37

So up to this point, we have looked at the standard form of the equation of a line, $Ax + By = C$, and the slope-intercept form of a line, $y = mx + b$. We now introduce the **point-slope form** of a line.

Note: He derives the following form of a line by using the formula for the slope of a line. Enjoy!

Point-Slope Form of a Line
An equation of a nonvertical line of slope m that contains the point (x_1, y_1) is

1. **Example:** Find an equation of the line whose slope is 5 and contains the point $(4, -3)$. Write the equation in slope-intercept form.

   Final answer: _____

2. **Example:** Find the equation of a horizontal line through the point $(3, -4)$.
   Note: Pay special attention to the alternative method he provides on how to find the equation of a horizontal line.

   Final answer: _____

Course: Name:
Instructor: Section:

Section 3.5 – Objective 2: Find the Equation of a Line Given Two Points
Video Length – 8:18

3. **Example:** Find an equation of the line that passes through $(-2, 1)$ and $(7, 4)$. Write the equation in slope-intercept form.

Write the steps in words	Show the steps with math
Step 1	
Step 2	
Step 3	
Graph the line.	

108 Copyright © 2014 Pearson Education, Inc.

Course: Name:
Instructor: Section:

4. **Example:** Find an equation of the line that passes through $(3,6)$ and $(3,-1)$.

Final answer: _____

Course:
Instructor:

Name:
Section:

Section 3.6 Video Guide
Parallel and Perpendicular Lines

Objectives:
1. Determine Whether Two Lines Are Parallel
2. Find the Equation of a Line Parallel to a Given Line
3. Determine Whether Two Lines Are Perpendicular
4. Find the Equation of a Line Perpendicular to a Given Line

Section 3.6 – Objective 1: Determine Whether Two Lines Are Parallel
Video Length – 4:17

Definition
When two lines (in the Cartesian plane) do not intersect, they are said to be _____. Two nonvertical lines are parallel if and only if their _____ are _____ and they have _____ ___-_____. Vertical lines are parallel if they have _____ ___-_____.

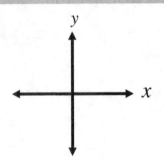

1. **Example:** Determine whether the lines are parallel.

$$L_1 : 6x + 2y = 9$$
$$L_2 : -3x - y = 3$$

Final answer: _____

Course: Name:
Instructor: Section:

Section 3.6 – Objective 2: Find the Equation of a Line Parallel to a Given Line
Video Length – 3:12

2. **Example:** Find the equation of the line parallel to $4x + y = -8$ and through the point $(2, -3)$.

Write the steps in words	Show the steps with math
Step 1	
Step 2	

Final answer: _____

Course: Name:
Instructor: Section:

Section 3.6 – Objective 3: Determine Whether Two Lines Are Perpendicular
Video Length – 4:17

Definition
When two lines intersect at a right angle $90°$, they are said to be _____. Two nonvertical lines are perpendicular if and only if the _____ of their _____ is _____. Any _____ line is perpendicular to any _____ line.

We can also say that two (*nonvertical*) lines are perpendicular if the slope of one of the lines is the negative reciprocal of the slope of the other line.

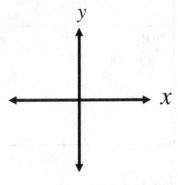

3. **Example:** Determine whether the lines are perpendicular.

$$L_1 : x + 3y = -15$$
$$L_2 : -3x + y = -1$$

Final answer: _____

Course: Name:
Instructor: Section:

Section 3.6 – Objective 4: Find the Equation of a Line Perpendicular to a Given Line
Video Length – 4:00

4. **Example:** Find an equation of the line perpendicular to $-2x + 5y = 3$ and through the point $(2, -3)$.

Write the steps in words	Show the steps with math
Step 1	
Step 2	
Step 3	

Final answer: _____

Course: Name:
Instructor: Section:

Section 3.7 Video Guide
Linear Inequalities in Two Variables

Objectives:
1. Determine Whether an Ordered Pair Is a Solution to a Linear Inequality
2. Graph Linear Inequalities
3. Solve Problems Involving Linear Inequalities

Section 3.7 – Objective 1: Determine Whether an Ordered Pair Is a Solution to a Linear Inequality

Video Length – 3:40

Definition

_____ _____ in _____ _____ are inequalities in one of the forms

where A and B are not both zero.

Definition

A linear inequality in two variables x and y is _____ by an ordered pair (a,b) if a _____ results when _____ is replaced by _____ and _____ is replaced by _____.

1. **Example:** Determine whether the ordered pair $(5,-1)$ is a solution to the inequality $4x-5y \geq 12$. How about $(3,1)$?

   Final answer: _____

Course: Name:
Instructor: Section:

Section 3.7 – Objective 2: Graph Linear Inequalities
Video Length – 17:28

Definition
A _____ of a _____ in _____ x and y consists of all points (x, y) whose coordinates _____ the inequality.

Let's graph the following linear inequality $3x + 4y \geq 12$.

Note: He graphs this linear inequality using two different methods.

Consider the graph of the inequality $y < -3x + 9$

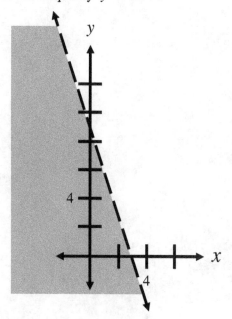

When graphing an inequality that is strict (_____ or _____) use a _____ line.

When graphing an inequalities that are non-strict (_____ or _____) use a _____ line.

Course:
Instructor:

Name:
Section:

2. **Example:** Graph the inequality $2x + 5y \geq 10$.

3. **Example:** Graph the inequality $2x - 3y < 0$

Course: Name:
Instructor: Section:

Section 3.7 – Objective 3: Solve Problems Involving Linear Inequalities
Video Length – 4:30

4. **Example:** Sharon gets $15 for allowance. She went to the store and bought gum that costs $1.50 a pack and candy bars that cost $2.00 each.

   (a) Write a linear inequality that describes Sharon's options for buying gum and candy bars with the $15.

   Final answer: _____

   (b) Can she buy 5 packs of gum and 6 candy bars?

   Final answer: _____

   (c) Can she buy 4 packs of gum and 4 candy bars?

   Final answer: _____

Course: Name:
Instructor: Section:

Course: Name:
Instructor: Section:

Section 4.1 Video Guide
Solving Systems of Linear Equations by Graphing

Objectives:
1. Determine If an Ordered Pair Is a Solution of a System of Linear Equations
2. Solve a System of Linear Equations by Graphing
3. Classify Systems of Linear Equations as Consistent or Inconsistent
4. Solve Applied Problems Involving Systems of Linear Equations

Section 4.1 – Objective 1: Determine If an Ordered Pair Is a Solution of a System of Linear Equations
Video Length – 6:10

Earlier, we learned that the graph of an equation, such as $2x + 5y = 10$, is the set of all (x, y) that make the equation a true statement. Now we will find ordered pairs, (x, y), that satisfy more than one equation at the same time.

Definition
A _____ of _____ _____ is a grouping of two or more linear equations where each equation contains one or more variables.

Examples of systems of linear equations:

Our goal is to find any and all ordered pairs that simultaneously satisfy both equations. We will learn a number of techniques to do this. However, the first thing we will learn is how to determine if an ordered pair is a solution.

Definition
A _____ of a system of equations consists of values of the _____ that satisfy _____ equation of the system. When we are solving a system of two linear equations containing two unknowns, we represent the solution as an _____ _____, (x, y).

1. **Example:** Determine whether $(-4, 16)$ is a solution to the system of the equations.

$$\begin{cases} y = -4x \\ y = -2x + 8 \end{cases}$$

Final answer: _____

Is $(-2, 8)$ a solution?

Course: Name:
Instructor: Section:

Section 4.1 – Objective 2: Solve a System of Linear Equations by Graphing
Part I – Text Examples 3 and 4
Video Length – 14:24

As mentioned before, there are a number of ways that can be used to solve a system of linear equations. The method that we will use in this section will focus on graphing. The bottom line is this:

The _____ of a linear equation represents the set of _____ _____ that make the

equation a _____ statement. So the _____ point of two linear equations will

represent the _____ to a system (since this point will make _____ equations true).

2. **Example:** Solve the following system by graphing: $\begin{cases} y = -3x - 6 \\ y = 2x - 1 \end{cases}$

Write the steps in words	*Show the steps with math*
Step 1	
Step 2	
Step 3	
Step 4	

       Final answer: _____

Course: Name:
Instructor: Section:

Note: Listen to his suggestion on the method used to graph the first equation.

3. **Example:** Solve the following system by graphing: $\begin{cases} 2x - 3y = -12 \\ 2x + 3y = 0 \end{cases}$

Final answer: _____

Course: Name:
Instructor: Section:

Section 4.1 – Objective 2: Solve a System of Linear Equations by Graphing
Part II – Text Examples 5 and 6
Video Length – 8:20

Sometimes when you graph two lines, the lines will not intersect at all. When this happens, you have the following:

Definition
A _____ _____ is a system of equations that has no solution.

4. **Example:** Solve by graphing: $\begin{cases} 3x - 2y = -4 \\ -9x + 6y = -1 \end{cases}$

Final answer: _____

So far, we have looked at two possibilities:

1) Lines _____ and the point of intersection represents the _____ to the system.

2) Lines are _____, so the system is _____. The solution set is the _____ _____.

The third possibility is that the line are coincident, meaning that they lie on top of each other. In this case, we call the system the following:

Definition
A _____ _____ is a system of equations where the solution set is the set of all points on the line.

Course: Name:
Instructor: Section:

5. **Example:** Solve by graphing: $\begin{cases} 3x + 2y = 2 \\ -6x - 4y = -4 \end{cases}$

Final answer: _____

Course:
Instructor:

Name:
Section:

Section 4.1 – Objective 3: Classify Systems of Linear Equations as Consistent or Inconsistent

Video Length – 7:20

We will now summarize the three different types of systems.

Graph	Number of Solutions	Type of System
Two lines _____ at one point.	If the lines _____, the system of equations has _____ solution given by the _____ of _____ .	_____ The equations are _____
_____ lines	If the lines are _____ , then the system of equations has _____ solution because the lines _____ _____ .	_____
Lines _____	If the lines _____ on top of each other, then the system has _____ many solutions. The solution set is the set of _____ _____ on the _____.	_____ The equations are _____ .

124   Copyright © 2014 Pearson Education, Inc.

Course: Name:
Instructor: Section:

Classifying a System of Equations Algebraically

Step 1: Write each equation in the system in _____-_____ form.

Step 2: (a) If the equations of the lines in the system have _____ _____, then

the lines will _____. The _____ of _____ represents the

_____. We say that the system is _____ and the equations are

_____.

    (b) If the equations of the lines have the _____ slope, but _____

_____-_____, then the lines are _____. We say that the system is

_____.

    (c) If the equations of lines have the _____ slope and the _____

_____-_____, then the lines are _____. We say that the

system is _____ and the equations are _____.

6. **Example:** Without graphing, determine the number of solutions of the system.

$$\begin{cases} y = -x + 5 \\ y = x - 5 \end{cases}$$

    Final answer: _____

7. **Example:** Without graphing, determine the number of solutions of the system.

$$\begin{cases} -x + 2y = 4 \\ 2x - 4y = -8 \end{cases}$$

    Final answer: _____

Course:
Instructor:
Name:
Section:

Section 4.1 – Objective 4: Solve Applied Problems Involving Systems of Linear Equations
Video Length – 4:50

8. **Example:** Amtech has two different cellular phone plans. Plan A charges a monthly fee of $14.95 plus $0.05 per minute. Plan B charges a monthly fee of $18.95 and $0.03 per minute. Determine the number of minutes for which the cost of the plans will be the same.

Final answer: _____

Note: Write your final answer as a complete sentence.

Course: Name:
Instructor: Section:

Section 4.2 Video Guide
Solving Systems of Linear Equations Using Substitution

Objectives:
1. Solve a System of Linear Equations Using the Substitution Method
2. Solve Applied Problems Involving Systems of Linear Equations

Section 4.2 – Objective 1: Solve a System of Linear Equations Using the Substitution Method
Part I – Text Examples 1, 2, and 3
Video Length – 19:34

In the last section we looked at solving systems of equations by graphing. This is a great tool for allowing us to visualize the solution to a system of equations. The problem, however, is if the intersection point is not made up of integer values. It would be pretty tough to figure out the exact coordinates. So we need to introduce some algebraic methods that will also allow us to solve a system of equations. The first algebraic method is called the **substitution method**.

1. **Example:** Solve the following system by substitution: $\begin{cases} 2x - y = 13 \\ -4x - 9y = 7 \end{cases}$

Write the steps in words	Show the steps with math
Step 1	
Step 2	
Step 3	
Step 4	
Step 5	

Final answer: _____

Course: Name:
Instructor: Section:

Steps for Solving a System of Two Linear Equations Containing Two Unknowns by Substitution

Step 1: _____ one of the equations for one of the unknowns.

Step 2: _____ the expression solved for in Step 1 into the _____ equation. The result will be a single linear equation in one unknown.

Step 3: _____ the linear equation in one unknown found in Step 2.

Step 4: _____ the value of the variable found in Step 3 into one of the _____ equations to find the value of the _____ variable.

Step 5: _____ your answer by substituting the ordered pair into _____ of the original equations.

2. **Example:** Solve the following system by substitution: $\begin{cases} x + 2y = -6 \\ \dfrac{2}{3}x + \dfrac{1}{3}y = -3 \end{cases}$

Final answer: _____

Course: Name:
Instructor: Section:

3. **Example:** Solve the following system by substitution: $\begin{cases} 2x + 4y = 11 \\ 3x + y = -6 \end{cases}$

Final answer: _____

Course: Name:
Instructor: Section:

Section 4.2 – Objective 1: Solve a System of Linear Equations Using the Substitution Method
Part II – Text Examples 4 and 5
Video Length – 6:55

Recall that there are three possibilities for a system of equations:

1. _____ and _____ (_____ solution).

2. _____ (_____ solution).

3. _____ and _____ (_____ many solutions).

4. **Example:** Solve the following system by substitution: $\begin{cases} 3x+6y=12 \\ x+2y=7 \end{cases}$

   Final answer: _____

5. **Example:** Solve the following system by substitution: $\begin{cases} x+y=1 \\ 3x+3y=3 \end{cases}$

   Final answer: _____

Course:
Instructor:
Name:
Section:

Section 4.2 – Objective 2: Solve Applied Problems Involving Systems of Linear Equations
Video Length – 3:12

6. **Example:** Wilson's Woodworks produce desk chairs that they sell to local home improvement stores. The cost y (in hundreds of dollars) to produce x desk chairs can be determined by the equation $y = 0.5x + 6$. The revenue y (in hundreds of dollars) from the sale of x desk chairs is given by the equation $y = 1.5x$. Find the break-even point.

Final answer: _____

Course:
Instructor:

Name:
Section:

Section 4.3 Video Guide
Solving Systems of Linear Equations Using Elimination

Objectives:
1. Solve a System of Linear Equations Using the Elimination Method
2. Solve Applied Problems Involving Systems of Linear Equations

Section 4.3 – Objective 1: Solve a System of Linear Equations Using the Elimination Method
Part I – Text Examples 1, 2, and 3
Video Length – 20:37

We now have two methods for solving a system of linear equations – the graphing method and the method of substitution. We will now introduce a third method for solving a system of linear equations called the **elimination method**.

1. **Example:** Solve the following system by elimination: $\begin{cases} 5x - 3y = 14 \\ 2x - y = 6 \end{cases}$

Write the steps in words	Show the steps with math
Step 1	
Step 2	
Step 3	
Step 4	
Step 5	

Final answer: _____

Course: Name:
Instructor: Section:

2. **Example:** Solve the following system by elimination: $\begin{cases} \dfrac{3}{2}x - \dfrac{y}{8} = -1 \\ 16x + 3y = -28 \end{cases}$

Final answer: _____

Steps for Solving a System of Two Linear Equations by Elimination

Step 1: Write each equation in _____, _____ .

Step 2: Make sure that the _____ on one of the variables are _____ _____ by _____ (or _____) both sides of one (or both) equation(s) by a nonzero constant.

Step 3: _____ the equations to eliminate the variable whose coefficients are now additive inverses. _____ the resulting equation for the remaining unknown.

Step 4: _____ the value of the variable found in Step 3 into one of the _____ equations to find the value of the _____ variable.

Step 5: _____ your answer by substituting the ordered pair into _____ of the original equations.

Course: Name:
Instructor: Section:

3. **Example:** Solve the following system by elimination: $\begin{cases} 2x = -5y - 11 \\ 3x + 2y = 11 \end{cases}$

Final answer: _____
Note: DON'T FORGET TO CHECK YOUR ANSWER!!!

Course:
Instructor:
Name:
Section:

Section 4.3 – Objective 1: Solve a System of Linear Equations Using the Elimination Method
Part II – Text Examples 4 and 5
Video Length – 7:22

Remember there are three types of systems – consistent/independent, inconsistent and consistent/dependent systems. We will now use the elimination method on an inconsistent system.

4. **Example:** Solve the following system by elimination: $\begin{cases} 5x - y = 3 \\ -10x + 2y = 2 \end{cases}$

Final answer: _____

Now let's use the elimination method on a consistent/dependent system.

5. **Example:** Solve the following system by elimination: $\begin{cases} 6x - 4y = 8 \\ -9x + 6y = -12 \end{cases}$

Final answer: _____

Course: Name:
Instructor: Section:

Section 4.3 – Objective 2: Solve Applied Problems Involving Systems of Linear Equations
Video Length – 3:46

6. **Example:** Findlay Community Center is holding their annual musical. 650 tickets were sold for a total value of $4375. If adult tickets cost $7.50, and student tickets cost $3.50, how many of each kind of ticket were sold?

   Let a represent the number of adult tickets and s the number of student tickets. Then we can determine how many adult tickets and how many student tickets were sold by solving the following system

   $$\begin{cases} a + s = 650 \\ 7.5a + 3.5s = 4375 \end{cases}$$

   Final answer: _____

Course: Name:
Instructor: Section:

Section 4.4 Video Guide
Solving Direct Translation, Geometry, and Uniform Motion Problems Using Systems of Linear Equations

Objectives:
1. Model and Solve Direct Translation Problems
2. Model and Solve Geometry Problems
3. Model and Solve Uniform Motion Problems

Section 4.4 – Objective 1: Model and Solve Direct Translation Problems
Video Length – 8:15

Earlier we developed mathematical models where the model ended up with one unknown in the equation or inequality. In this chapter, we will focus in on models that lead to two unknowns.

1. **Example:** Find two numbers whose sum is 56 and when three times the first is subtracted from the second, the difference is 4.

Write the steps in words	Show the steps with math
Step 1	
Step 2	
Step 3	
Step 4	
Step 5	
Step 6	

Course: Name:
Instructor: Section:

Section 4.4 – Objective 2: Model and Solve Geometry Problems
Part I – Text Example 2
Video Length – 8:03

2. **Example:** The perimeter of a rectangle field is 142 feet. If the length of the field is 15 feet longer than the width, find the dimensions of the field.

Final answer: _____

Course: Name:
Instructor: Section:

Section 4.4 – Objective 2: Model and Solve Geometry Problems
Part II – Text Example 3
Video Length – 7:53

Definition

Two angles are _____ if their sum is _____.

Two angles are _____ if their sum is _____.

Note: He admits to using "briefer" language in the definitions above. He then defines the two types of angles accordingly – two angles are complementary if the sum of the measure of their angles is $90°$ and two angles are supplementary if the sum of the measure of their angles is $180°$.

3. **Example:** Find the measure of two angles that are supplementary such that the measure of the larger angle is $32°$ greater than the measure of the smaller angle.

Final answer: _____

Course:
Instructor:
Name:
Section:

Section 4.4 – Objective 3: Model and Solve Uniform Motion Problems
Video Length – 13:09

4. **Example:** Suppose a plane is flying west a distance of 600 miles takes 6 hours. The return trip takes 5 hours. Find the airspeed of the plane and the effect of wind resistance on the plane.

Final answer: _____

Course: Name:
Instructor: Section:

Section 4.5 Video Guide
Solving Mixture Problems Using Systems of Linear Equations

Objectives:
1. Draw Up a Plan for Solving Mixture Problems
2. Set Up and Solve Money Mixture Problems
3. Set Up and Solve Dry Mixture and Percent Mixture Problems

Section 4.5 – Objective 1: Draw Up a Plan for Modeling Mixture Problems
Video Length – 5:18

Problems that involve mixing two or more substances are called _____

_____.

This type of problem can be set up using the model

_____ · _____ = _____

For example,

$$\underline{\text{the number of adults}} \cdot \underline{\text{the price of the adult ticket}} = \underline{\text{total revenue}}$$

or

$$\underline{\text{amount invested in treasury bonds}} \cdot \underline{\text{interest rate earned}} = \underline{\text{total interest}}$$

A table is helpful for organizing the given information.

Note: For the following problem, you will set up a table (like the one you drew above) and the system of equations that will solve the problem. However, you will not actually solve the problem.

1. **Example:** A theatre sells adult tickets for $12 and children tickets for $8. If the theatre sold 90 tickets and collected $920 in revenue, how many tickets were sold? How many children tickets were sold?

Course:
Instructor:
Name:
Section:

Section 4.5 – Objective 2: Set Up and Solve Money Mixture Problems
Video Length – 6:37

2. **Example:** Matthew received an $8000 Christmas bonus from his employer. He invested part of the bonus in a savings account earning 3% simple interest per year and the rest in a certificate of deposit (CD) that earns 4.5% simple interest annually. At the end of the year, he will receive $330 interest on his investments. How much money did Matthew invest in each account?

Final answer: _____

Course:
Instructor:
Name:
Section:

Section 4.5 – Objective 3: Set Up and Solve Dry Mixture and Percent Mixture Problems
Part I – Text Example 4
Video Length – 10:50

Note: Towards the end of the video, he checks (without computation) the reasonableness of his final answer. When it comes to problem solving, this is an important skill to have.

3. **Example:** A nut company wants to produce 20 pounds of nuts worth $4.50 per pound. To obtain this, they want to mix cashews worth $5.00 per pound with peanuts worth $3.00 per pound. How many pounds of each type should be used to make the desired 20 pounds worth $4.50 per pound.

Final answer: _____
Note: Don't forget to write your answer as a complete sentence.

Course: Name:
Instructor: Section:

Section 4.5 – Objective 3: Set Up and Solve Dry Mixture and Percent Mixture Problems
Part II – Text Example 5
Video Length – 10:26

4. **Example:** You work in the chemistry stockroom and your instructor has asked you to prepare 4 liters of 15% hydrochloric acid (HCl). Looking through the supply room you see that there is a bottle of 12% HCl and another 20% HCl. How much of each should you mix so that your instructor has the required solution?

Final answer: _____

Course: Name:
Instructor: Section:

Section 4.6 Video Guide
Systems of Linear Inequalities

Objectives:
1. Determine Whether an Ordered Pair Is a Solution of a System of Linear Inequalities
2. Graph a System of Linear Inequalities
3. Solve Applied Problems Involving Systems of Linear Inequalities

Section 4.6 – Objective 1: Determine Whether an Ordered Pair Is a Solution of a System of Linear Inequalities
Video Length – 3:01

Definition
An ordered pair _____ a system of linear inequalities if it makes _____ inequality in the system a true statement.

1. **Example:** (a) Determine if the ordered pair $(-5, 2)$ is a solution to the system of linear inequalities.

    (b) Determine if the ordered pair $(3, -2)$ is a solution to the system of linear inequalities.

$$\begin{cases} 3x + 2y \leq 5 \\ -x + 4y \leq 12 \end{cases}$$

Note: Answer each using a complete sentence. For example you can write the following: "The ordered pair, (a,b), is not a solution to the system of linear inequalities."

(a) **Final answer:** _____

(b) **Final answer:** _____

Copyright © 2014 Pearson Education, Inc.

Course:
Instructor:
Name:
Section:

Section 4.6 – Objective 2: Graph a System of Linear Inequalities
Video Length – 13:09

Now we will graph the solution of a system of linear inequalities. When graphing inequalities, recall the following:

Graph the inequality as a _____ line when the inequality is strict (_____ or _____).

Graph the inequality as a _____ line when the inequality is non-strict (_____ or _____).

We went over two ways to graph these inequalities. How you graph the inequalities is up to you. Essentially we will graph each inequality by itself and then find the overlapping region. This overlapping region represents the solution set.

2. **Example:** Graph the solution of the system: $\begin{cases} 3x + y < 9 \\ 2x + 5y \geq 10 \end{cases}$

The solution will be the set of all points that satisfy both of the inequalities in the system.

Note: Make sure YOU graph the other inequality in the system. Also, clearly mark the region that represents the solution set.

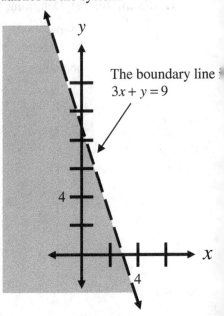

The boundary line $3x + y = 9$

3. **Example:** Graph the solution of the system: $\begin{cases} 3x + y \leq 5 \\ -2x + y \geq -10 \end{cases}$

Section 4.6 – Objective 2: Solve Applied Problems Involving Systems of Linear Inequalities
Video Length – 7:19

4. **Example:** Jacob recently retired and has up to $70,000 to invest. His financial planner has recommended that he place no more than $45,000 in corporate bonds and at least $20,000 in Treasury notes. A system of linear inequalities that models this situation is given by

$$\begin{cases} c + t \leq 70{,}000 \\ c \leq 45{,}000 \\ t \geq 20{,}000 \end{cases}$$

where c represents the amount invested in corporate bonds and t represents the amount invested in Treasury notes.

(a) Graph the system.
(b) Can Jacob put $25,000 in corporate bonds and $35,000 in Treasury notes?
(c) Can Jacob put $40,000 in corporate bonds and $15,000 in Treasury notes?

(a) **Final answer:**

(b) **Final answer:** _____

(c) **Final answer:** _____

Course: Name:
Instructor: Section:

Course: Name:
Instructor: Section:

Section 5.1 Video Guide
Adding and Subtracting Polynomials

Objectives:
1. Define Monomial and Determine the Degree of a Monomial
2. Define Polynomial and Determine the Degree of a Polynomial
3. Simplify Polynomials by Combining Like Terms
4. Evaluate Polynomials

Section 5.1 – Objective 1: Define Monomial and Determine the Degree of a Monomial
Video Length – 3:56

Definition
A _____ in _____ _____ is the product of a constant and a variable raised to a nonnegative power. A monomial in one variable is of the form

where a is a constant, x is a variable, and $k \geq 0$ is an _____. The constant a is called the _____ of the monomial. If $a \neq 0$, then k is called the _____ of the monomial.

So what happens if $a = 0$? _____

_____

                 Monomial   Coefficient   Degree

Examples of expressions that are NOT monomials:

$5a^{\frac{2}{3}}$ is not a monomial because _____

_____

$3b^{-4}$ is not a monomial because _____

_____

Definition
The _____ of the _____, _____ is the sum of the exponents, _____.

$2a^3$ $4y^3z^5$ $5a^3bc^2$

Course: Name:
Instructor: Section:

Section 5.1 – Objective 2: Define Polynomial and Determine the Degree of a Polynomial
Part I – Define Polynomials
Video Length – 3:24

Definition
A _____ is a monomial or the _____ of monomials.

Examples of polynomials: Note: *Pay attention to the number of terms in each polynomial.*

Definition
A polynomial is in _____ _____ if it is written with the terms in _____ order according to degree.

Definition
The _____ of a polynomial is the _____ _____ of all the terms of the polynomial.

Polynomial   Degree

Examples of algebraic expressions that are NOT polynomials:

$3x^{-2} - 6x + 2 \quad 5a^{\frac{2}{3}}$ is not a polynomial because _____

_____

$\dfrac{3}{p^2}$ is not a polynomial because _____

_____

$\dfrac{2ab-3}{a+7}$ is not a polynomial because _____

_____

Course: Name:
Instructor: Section:

Definition
A polynomial that has _____ monomials that are not like terms is called a _____.

Examples of binomials:

Definition
A polynomial that contains _____ monomials that are not like terms is called a
_____ .

Examples of trinomials:

Course:
Instructor:
Name:
Section:

Section 5.1 – Objective 3: Simplify Polynomials by Combining Like Terms
Part I – Add Polynomials
Video Length – 4:24

To add polynomials, combine the like terms of the polynomials.

Note: He does the following problem using both "horizontal addition" and "vertical addition."

1. **Example:** Simplify: $(5x^2 - x + 6) + (2x^2 - 3)$

   Final answer: $(5x^2 - x + 6) + (2x^2 - 3) = $ _____

2. **Example:** Simplify: $(5x^2y - 3xy + 12xy^2) + (x^2y + 12xy - 5xy^2)$

   Final answer: $(5x^2y - 3xy + 12xy^2) + (x^2y + 12xy - 5xy^2) = $ _____

Course: Name:
Instructor: Section:

Section 5.1 – Objective 3: Simplify Polynomials by Combining Like Terms
Part II – Subtract Polynomials
Video Length – 3:24

To subtract one polynomial from another, add the opposite of each term in the polynomial following the subtraction sign and then combine like terms.

Note: He does the following problem using both "horizontal subtraction" and "vertical subtraction."

3. **Example:** Subtract: $(3a^4 + 5a - 6) - (2a^4 + 2a - 3)$

   Final answer: $(3a^4 + 5a - 6) - (2a^4 + 2a - 3) =$ _____

Later on down the road, vertical subtraction will be used. So it would be a good idea to practice this method.

Course:
Instructor:

Name:
Section:

Section 5.1 – Objective 4: Evaluate Polynomials
Video Length – 3:50

Definition
To _____ a polynomial, we _____ the given number for the value of the variable and simplify.

4. **Example:** Evaluate the polynomial $3x^3 - 2x^2 + x - 1$ for

(a) $x = 2$

(b) $x = -4$

Final answer: _____

Final answer: _____

Course: Name:
Instructor: Section:

Section 5.2 Video Guide
Multiplying Monomials: The Product and Power Rules

Objectives:
1. Simplify Exponential Expressions Using the Product Rule
2. Simplify Exponential Expressions Using the Power Rule
3. Simplify Exponential Expressions Containing Products
4. Multiply a Monomial by a Monomial

Section 5.2 – Objective 1: Simplify Exponential Expressions Using the Product Rule
Video Length – 8:25

Recall the definition for raising a real number to a positive integer exponent:

For example, suppose we have 5^4. What does this mean in expanded form?

So what we're going to do now is develop some rules that we can use for exponents. For example, suppose you are asked to simplify $x^3 \cdot x^4$:

Product Rule for Exponents
If a is a real number and m and n are natural numbers, then

$$\underline{} \cdot \underline{} = \underline{}$$

In other words, when you have _____

_____

1. **Example:** Evaluate each expression:

    (a) $3^2 \cdot 3^3$ (a) **Final answer:** $3^2 \cdot 3^3 = $ _____

    (b) $(-2)^3 \cdot (-2)^2$ (b) **Final answer:** $(-2)^3 \cdot (-2)^2 = $ _____

    (c) $n^3 \cdot n^5$ (c) **Final answer:** $n^3 \cdot n^5 = $ _____

    (d) $a \cdot a^2 \cdot b^3$ (d) **Final answer:** $a \cdot a^2 \cdot b^3 = $ _____

Course:
Instructor:
Name:
Section:

Section 5.2 – Objective 2: Simplify Exponential Expressions Using the Power Rule
Video Length – 4:08

Now will take an exponential expression and raise it to an exponent. For example,

$$\left(2^3\right)^4 =$$

Power Rule for Exponents
If a is a real number and m and n are natural numbers, then

$$\underline{\qquad} = \underline{\qquad}$$

2. **Example:** Simplify each expression. Write the answer in exponential form.

   (a) $\left(3^2\right)^4$ (a) **Final answer:** $\left(3^2\right)^4 = \underline{\qquad}$

   (b) $\left[(-2)^2\right]^3$ (b) **Final answer:** $\left[(-2)^2\right]^3 = \underline{\qquad}$

   (c) $\left[(-x)^3\right]^7$ (c) **Final answer:** $\left[(-x)^3\right]^7 = \underline{\qquad}$

Course: Name:
Instructor: Section:

Section 5.2 – Objective 3: Simplifying Exponential Expressions Containing Products
Video Length – 5:10

Consider the following:

$$(x \cdot y)^4 =$$

Product to a Power Rule for Exponents
If a and b are real numbers and n is a natural number, then

$$\underline{} = \underline{}$$

3. **Example:** Simplify each expression:

(a) $(3c^2)^2$ (a) **Final answer:** $(3c^2)^2 = $ _____

(b) $(-2y^5)^3$ (b) **Final answer:** $(-2y^5)^3 = $ _____

Below is a summary for all the rules for multiplying monomials that we just developed.

Rules for Multiplying Monomials

Product Rule for Exponents
If a is a real number and m and n are natural numbers, then

$$\underline{} \cdot \underline{} = \underline{}$$

Power Rule for Exponents
If a is a real number and m and n are natural numbers, then

$$\underline{} = \underline{}$$

Product to a Power Rule for Exponents
If a and b are real numbers and n is a natural number, then

$$\underline{} = \underline{}$$

Course: Name:
Instructor: Section:

Section 5.2 – Objective 4: Multiply a Monomial by a Monomial
Video Length – 2:59

We will now make use of the rules that we just learned to multiply two monomials.

4. **Example:** Multiply and simplify:

   (a) $(5x^2)(6x^4)$ (a) **Final answer:** $(5x^2)(6x^4) = $ _____

   (b) $(5z^2) \cdot (-4z^7)$ (b) **Final answer:** $(5z^2) \cdot (-4z^7) = $ _____

   (c) $(-4a^7b) \cdot (-6a^4b^5)$ (c) **Final answer:** $(-4a^7b) \cdot (-6a^4b^5) = $ _____

Course: Name:
Instructor: Section:

Section 5.3 Video Guide
Multiplying Polynomials

Objectives:
1. Multiply a Polynomial by a Monomial
2. Multiply Two Binomials Using the Distributive Property
3. Multiply Two Binomials Using the FOIL Method
4. Multiply the Sum and Difference of Two Terms
5. Square a Binomial
6. Multiply a Polynomial by a Polynomial

Section 5.3 – Objective 1: Multiply a Polynomial by a Monomial
Video Length – 5:00

Recall the Distributive Property:

$$a(b+c) =$$

When we multiply a polynomial by a monomial, we use the *Extended Form of the Distributive Property*.

Extended Form of the Distributive Property

$$a(b+c+\cdots+z) =$$

where $a, b, c, \ldots, z$ are real numbers.

1. **Example:** Use the distributive property to multiply and simplify each of the following:

   (e) $4x^3(x^2 - 3x + 2)$ (a) **Final answer:** $4x^3(x^2 - 3x + 2) =$ _____

   (f) $-4ab(2a - 3ab + 5a^2)$ (b) **Final answer:** $-4ab(2a - 3ab + 5a^2) =$ _____

Copyright © 2014 Pearson Education, Inc.

Course: Name:
Instructor: Section:

Section 5.3 – Objective 2: Multiply Two Binomials Using the Distributive Property
Video Length – 9:48

Now we will discuss how to use the Distributive Property to multiply two binomials. Remember a binomial is a polynomial that contains two terms. There will be two approaches that we will take to multiply binomials. The first will utilize the Distributive Property.

Two multiply two binomials, use the Distributive Property by distributing the first binomial to each term in the second binomial.

2. **Example:** Multiply: $(7x+3)(2x+5)$

   Final answer: $(7x+3)(2x+5) = $ _____

3. **Example:** Multiply: $(2x-3)(3x+7)$

   Final answer: $(2x-3)(3x+7) = $ _____

Course: Name:
Instructor: Section:

Section 5.3 – Objective 3: Multiply Two Binomials Using the FOIL Method
Video Length – 6:36

The acronym FOIL stands for _____, _____, _____, _____.

$$(a+b)(c+d)$$

For example, find the product, $(x+2)(x+6)$, using the FOIL method.

$$(x+2)(x+6)$$

4. **Example:** Multiply: $(2z+3)(z-5)$

   Final answer: $(2z+3)(z-5) = $ _____

5. **Example:** Multiply: $(3a+4b)(5a-b)$

   Final answer: $(3a+4b)(5a-b) = $ _____

Course:
Instructor:
Name:
Section:

Section 5.3 – Objective 4: Multiply The Sum and Difference of Two Terms
Video Length – 5:55

Certain products are called **special products** because they occur quite frequently and result in the same pattern.

Let's first consider the product that is referred to as the difference of two squares.

6. **Example:** Find the product: $(n-4)(n+4)$

   Final answer: $(n-4)(n+4) = $ _____

Product of the Sum and Difference of Two Terms

$$(a-b)(a+b) = \underline{}$$

7. **Example:** Find each product:

   (a) $(5x+3)(5x-3)$ (a) **Final answer:** $(5x+3)(5x-3) = $ _____

   (b) $(8x+2y)(8x-2y)$ (b) **Final answer:** $(8x+2y)(8x-2y) = $ _____

Course: Name:
Instructor: Section:

Section 5.3 – Objective 5: Square a Binomial
Video Length – 8:46

Note: He points out a common error that students frequently make. Pay close attention to this error. If you make this error on an assignment, quiz, or test, you will give your instructor a heart attack!!!

8. **Example:** Find the product: $(x+6)^2$

Final answer: $(x+6)^2 = $ _____

Note: After he finishes this example, he develops a formula for the square of a binomial (in red ink). Write the formulas in the space below.

Squares of Binomials

9. **Example:** Find the product: $(7z-2)^2$

Final answer: $(7z-2)^2 = $ _____

Course: Name:
Instructor: Section:

Section 5.3 – Objective 6: Multiply a Polynomial by a Polynomial
Video Length – 2:05

To multiply any two polynomials, we make repeated use of the Distributive Property.

10. Example: Find the product: $(w-1)(2w^2+7w+3)$

Final answer: $(w-1)(2w^2+7w+3) = $ _____

Course: Name:
Instructor: Section:

Section 5.4 Video Guide
Dividing Monomials: The Quotient Rule and Integer Exponents

Objectives:
1. Simplify Exponential Expressions Using the Quotient Rule
2. Simplify Exponential Expressions Using the Quotient to a Power Rule
3. Simplify Exponential Expressions Using Zero as an Exponent
4. Simplify Exponential Expressions Involving Negative Exponents
5. Simplify Exponential Expressions Using the Laws of Exponents

Section 5.4 – Objective 1: Simplify Exponential Expressions Using the Quotient Rule
Video Length – 6:11

Consider the following example:

$$\frac{x^5}{x^2} =$$

The Quotient Rule for Exponents
If a is a real number and if m and n are positive integers, then

$$\underline{\hspace{2cm}} = \underline{\hspace{2cm}} \quad \text{if } a \neq 0$$

In other words, when you have the _____

_____

1. **Example:** Simplify each expression.

    (a) $\dfrac{3^6}{3^2}$ (b) $\dfrac{y^7}{y^4}$

    Final answer: $\dfrac{3^6}{3^2} = $ _____   **Final answer:** $\dfrac{y^7}{y^4} = $ _____

2. **Example:** Simplify each expression.

    (a) $\dfrac{16z^5}{12z^3}$ (b) $\dfrac{-35a^7b^3}{25a^6b}$

    Final answer: $\dfrac{16z^5}{12z^3} = $ _____   **Final answer:** $\dfrac{-35a^7b^3}{25a^6b} = $ _____

Course: Name:
Instructor: Section:

Section 5.4 – Objective 2: Simplify Exponential Expressions Using the Quotient to a Power Rule

Video Length – 5:27

Consider the following scenario:

$$\left(\frac{2}{3}\right)^4$$

Quotient to a Power Rule for Exponents

$$\underline{} = \underline{} \quad \text{if } b \neq 0.$$

3. **Example:** Simplify each expression:

   (a) $\left(\dfrac{x}{y}\right)^{10}$

   Final answer: $\left(\dfrac{x}{y}\right)^{10} = \underline{}$

   (b) $\left(\dfrac{3a^3}{b^2}\right)^4$

   Final answer: $\left(\dfrac{3a^3}{b^2}\right)^4 = \underline{}$

4. **Example:** Simplify the expression: $\left(\dfrac{-3x^2}{y}\right)^2$

   Final answer: $\left(\dfrac{-3x^2}{y}\right)^2 = \underline{}$

Course: Name:
Instructor: Section:

Section 5.4 – Objective 3: Simplify Exponential Expressions Using Zero as an Exponent
Video Length – 4:21

Up to this point we've only considered expressions where the exponents are natural numbers (i.e. 1, 2, 3, 4, 5, etc.). We will now extend our definitions so they apply to integers. The first value we need to consider beyond the natural numbers is zero.

If a is a nonzero real number (that is $a \neq 0$), we define

$$_____ = _____$$

5. **Example:** Simplify each expression.

(a) $(24ab)^0$

Final answer: $(24ab)^0 = $ _____

(b) $\left(\dfrac{3a^3}{b^2}\right)^0$

Final answer: $\left(\dfrac{3a^3}{b^2}\right)^0 = $ _____

Course: Name:
Instructor: Section:

Section 5.4 – Objective 4: Simplify Exponential Expressions Involving Negative Exponents
Part I – Text Examples 6, 7, 8, and 9
Video Length – 8:28

Now we will look at negative integer exponents. Suppose we have the following:

$$\frac{y^3}{y^5}$$

If n is a positive integer and if a is a nonzero real number (that is, $a \neq 0$), then we define

$$\underline{\hspace{1in}} = \underline{\hspace{1in}}$$

6. **Example:** Simplify each expression.

   (a) $(-4)^{-2}$ (a) **Final answer:** $(-4)^{-2} = $ _____

   (b) $(ab)^{-3}$ (b) **Final answer:** $(ab)^{-3} = $ _____

Now suppose we have the expression:

$$-4^{-2}$$

This expression can be read as "take the _____."
Notice how -4^{-2} differs from $(-4)^{-2}$.

Note: He suggests saying the expression out loud in words. This really does help because it makes you more aware of the order of operations, especially when parentheses are involved.

7. **Example:** Simplify each expression.

   (a) $4z^{-3}$ (a) **Final answer:** $4z^{-3} = $ _____

   (b) $(4z)^{-3}$ (b) **Final answer:** $(4z)^{-3} = $ _____

From the previous example, we can see how the use of parentheses makes for a different problem. SO BE CAREFUL!!!

Course:
Instructor:

Name:
Section:

Section 5.4 – Objective 4: Simplify Exponential Expressions Involving Negative Exponents
Part II – Text Examples 10, 11, and 12
Video Length – 6:56

Note: A student asks, "will we ever have negative integers in the denominator?" What she meant to say was, "will we ever have negative integer exponents in the denominator?"

In other words, will we have something like $\dfrac{1}{a^{-n}}$?

8. **Example:** Simplify each expression.

   (a) $\dfrac{1}{x^{-3}}$

   (a) **Final answer:** $\dfrac{1}{x^{-3}} = $ _____

   (b) $\dfrac{1}{5x^{-3}}$

   (b) **Final answer:** $\dfrac{1}{5x^{-3}} = $ _____

   (c) $\dfrac{1}{(5x)^{-3}}$

   (c) **Final answer:** $\dfrac{1}{(5x)^{-3}} = $ _____

Again, note how the use of parentheses plays a big role in the value of the expression.

Course:
Instructor:

Name:
Section:

Note: He does the following example in two ways. A "long winded" way and a "short cut" way.

9. **Example:** Simplify the expression: $\left(\dfrac{2}{3}\right)^{-2}$

Final answer: $\left(\dfrac{2}{3}\right)^{-2} =$ _____

10. **Example:** Simplify the expression: $\left(-\dfrac{y^2}{5}\right)^{-2}$

Final answer: $\left(-\dfrac{y^2}{5}\right)^{-2} =$ _____

Course: Name:
Instructor: Section:

Section 5.4 – Objective 5: Simplify Exponential Expressions Using the Laws of Exponents
Part I – Text Examples 13 and 14
Video Length – 8:21

We will now put together all the different rules for simplifying exponents. To _____ an exponent means to perform all indicated operations (multiplication/division/power to power) AND to _____ _____ _____ .

11. **Example:** Simplify the expression: $\left(\dfrac{4}{3}x^2 y^{-3}\right)\left(-9x^{-4} y^5\right)$

Write the steps in words	Show the steps with math
Step 1	
Step 2	
Step 3	
Note: Remember, NOOOO NEGATIVE EXPONENTS!!!	

Final answer: $\left(\dfrac{4}{3}x^2 y^{-3}\right)\left(-9x^{-4} y^5\right) = $ _____

Course:
Instructor:

Name:
Section:

12. Example: Simplify the expression: $\dfrac{-25a^5b^{-1}}{15a^{-2}b^2}$

Final answer: $\dfrac{-25a^5b^{-1}}{15a^{-2}b^2} = $ _____

Course: Name:
Instructor: Section:

Section 5.4 – Objective 5: Simplify Exponential Expressions Using the Laws of Exponents
Part II – Text Example 15
Video Length – 7:55

Note: There are a lot of steps involved with this one so please be patient. Also, make sure your work is neat and organized.

13. **Example:** Simplify the expression: $\left(5a^{-2}b\right)\left(\dfrac{2a^{-2}b^3}{3ab}\right)^{-2}$

Final answer: $\left(5a^{-2}b\right)\left(\dfrac{2a^{-2}b^3}{3ab}\right)^{-2} = $ _____

Copyright © 2014 Pearson Education, Inc.   173

Course: Name:
Instructor: Section:

Section 5.5 Video Guide
Dividing Polynomials

Objectives:
1. Divide a Polynomial by a Monomial
2. Divide a Polynomial by a Binomial

Section 5.5 – Objective 1: Divide a Polynomial by a Monomial
Video Length – 10:31

So far in this chapter, we have learned how to add, subtract, and multiply polynomials. The only operation that remains is division. We will start by dividing a polynomial by a monomial.

1. **Example:** Divide and simplify: $\dfrac{10a^3 - 25a}{5a}$

    Final answer: $\dfrac{10a^3 - 25a}{5a} = $ _____

2. **Example:** Divide and simplify: $\dfrac{10t^4 - 35t^3 + 5t^2}{5t^2}$

    Final answer: $\dfrac{10t^4 - 35t^3 + 5t^2}{5t^2} = $ _____

Course:
Instructor:

Name:
Section:

Section 5.5 – Objective 2: Divide a Polynomial by a Binomial
Part I – Text Example 4
Video Length – 4:39

Now we are going to divide a polynomial by a binomial. This uses long division. Suppose you are asked to compute $\dfrac{328}{12}$:

Course: Name:
Instructor: Section:

Section 5.5 – Objective 2: Divide a Polynomial by a Binomial
Part II – Text Example 5
Video Length – 14:13

3. **Example:** Divide: $\dfrac{c^2 + 3c - 2}{c + 1}$

Write the steps in words	Show the steps with math
Step 1	
Step 2	
Step 3	
Step 4 Note: He mentions that you can stop dividing when the degree of the "new dividend" is less than the degree of the divisor. Remember, this "new dividend" is also known as the remainder.	
Step 5	

Final answer: $\dfrac{c^2 + 3c - 2}{c + 1} = $ _____

Course:
Instructor:

Name:
Section:

4. **Example:** Divide: $(y^2 - 5y + 6) \div (y - 2)$

Final answer: $(y^2 - 5y + 6) \div (y - 2) = $ _____

Course:
Instructor:

Name:
Section:

Section 5.5 – Objective 2: Divide a Polynomial by a Binomial
Part III – Text Example 6
Video Length – 4:53

In the last two examples, the divisor had a coefficient of one on the variable and the exponent on the variable was also one. In this problem, the coefficient on the divisor is two. But the approach that we take is the same.

5. **Example:** Divide: $\dfrac{4x^2 - 4x - 9}{2x + 3}$

Final answer: $\dfrac{4x^2 - 4x - 9}{2x + 3} =$ _____

Course: Name:
Instructor: Section:

Section 5.5 – Objective 2: Divide a Polynomial by a Binomial
Part IV – Text Example 7
Video Length – 5:04

6. **Example:** Divide: $\dfrac{7 + 2b^4 + 5b^2}{b^2 + 2}$

   Note: Listen carefully to what he says about how you should write the dividend.

   Final answer: $\dfrac{7 + 2b^2 + 5b^2}{b^2 + 2} =$ _____

Course: Name:
Instructor: Section:

Section 5.6 Video Guide
Applying Exponent Rules: Scientific Notation

Objectives:
1. Convert Decimal Notation to Scientific Notation
2. Convert Scientific Notation to Decimal Notation
3. Use Scientific Notation to Multiply and Divide

Section 5.6 – Objective 1: Convert Decimal Notation to Scientific Notation
Video Length – 8:28

When numbers are incredibly large or incredibly small, it is often the case that folks will express the numbers using scientific notation rather than decimal notation. Why? Because decimal notation is hard to read.

Definition
When a number has been written as the product of a number x, where ___ $\leq x <$ ___ , and a power of

_____, it is said to be in _____ _____. That is, a number is written in

scientific notation when it is in the form

where

For example, consider
$$8200 =$$

Steps to Convert from Decimal Notation to Scientific Notation

To change a positive number into scientific notation:

Step 1: Count the number N of decimal places that the decimal point must be moved in order to

arrive at a number x, where ___ $\leq x <$ ___ .

Step 2: If the original number is greater than or equal to 1, the scientific notation is ___×_____ . If

the original number is between 0 and 1, the scientific notation is ___×_____ .

Course: Name:
Instructor: Section:

1. **Example:** Write 67,300 in scientific notation.

    Final answer: 67,300 = _____

2. **Example:** Write 0.000000086 in scientific notation.

    Final answer: 0.000000086 = _____

Course: Name:
Instructor: Section:

Section 5.6 – Objective 2: Convert Scientific Notation to Decimal Notation
Video Length – 6:54

Now we will convert from scientific notation to decimal notation.

Steps to Convert from Scientific Notation to Decimal Notation

To change a positive number into scientific notation:

Step 1: Determine the exponent on the number 10.

Step 2: If the exponent is _____, then move the decimal _____ places to the

_____ . If the exponent is _____, then move the decimal place _____

places to the _____ . Add zeros, as needed.

---

3. **Example:** Write 9.1×10^4 in decimal notation.

   Final answer: $9.1 \times 10^4 = $ _____

Course: Name:
Instructor: Section:

4. **Example:** Write 6.72×10^{-3} in decimal notation.

Final answer: 6.72×10^{-3} = _____

Course: Name:
Instructor: Section:

Section 5.6 – Objective 3: Use Scientific Notation to Multiply and Divide
Video Length – 6:49

Now we will multiply and divide using scientific notation. The following two rules will be used for this multiplication and division:

5. **Example:** Multiply: $(3 \times 10^{-4}) \cdot (8 \times 10^{-5})$
   Express the answer in scientific notation.

   Final answer: $(3 \times 10^{-4}) \cdot (8 \times 10^{-5}) =$ _____

6. **Example:** Divide: $\dfrac{4.8 \times 10^7}{1.2 \times 10^2}$
   Express the answer in scientific notation.

   Final answer: $\dfrac{4.8 \times 10^7}{1.2 \times 10^2} =$ _____

Course: Name:
Instructor: Section:

Section 6.1 Video Guide
Greatest Common Factor and Factoring by Grouping

Objectives:
1. Find the Greatest Common Factor of Two or More Expressions
2. Factor Out the Greatest Common Factor in Polynomials
3. Factor Polynomials by Grouping

Section 6.1 – Introduction
Video Length – 1:37

In the last chapter we learned how to multiply two expressions. Now we're going to reverse that process. Remember, whenever we do something in math, we always want to have the ability to undo it.

Think of _____ as the "_____" of _____.

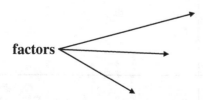

Definition
The expressions on the left side are called _____ of the expression on the right side.

To factor a polynomial means to write the polynomial as a _____ of _____ or _____ polynomials.

Basically, _____ is _____ the process of _____.

Course: Name:
Instructor: Section:

Section 6.1 – Objective 1: Find the Greatest Common Factor of Two or More Expressions
Part I – Text Examples 1 and 2
Video Length – 8:19

Definition
The _____ _____ _____ (_____) of a list of algebraic expressions is the largest expression that divides evenly into all the expressions.

1. **Example:** Find the GCF of 16 and 24.

Write the steps in words	Show the steps with math
Step 1	
Step 2	
Step 3	

   Final answer: GCF = _____

Steps to Find the Greatest Common Factor of a List of Numbers

Step 1: Write each number as a _____ of _____ _____ .

Step 2: Determine the _____ _____ _____ .

Step 3: Find the _____ of the _____ _____ found in Step 2. This number is the _____ .

2. **Example:** Find the GCF of 60, 75, and 135.

   Final answer: GCF = _____

186   Copyright © 2014 Pearson Education, Inc.

Course: Name:
Instructor: Section:

Section 6.1 – Objective 1: Find the Greatest Common Factor of Two or More Expressions
Part II
Video Length – 4:19

3. **Example:** Find the GCF of x^3, x^4 and x^5.

   Final answer: GCF = _____
   Note: If you have the same variable expression base, the smallest exponent that the variable factor is raised to gives rise to the GCF.

4. **Example:** Find the GCF of $12n^3z^2$ and $18n^2z^5$.

   Final answer: GCF = _____

5. **Example:** Find the GCF of $6(x-y)$ and $15(x-y)^3$.

   Final answer: GCF = _____

Course: Name:
Instructor: Section:

Section 6.1 – Objective 2: Factor Out the Greatest Common Factor in Polynomials
Part I – Text Examples 6, 7, and 8
Video Length – 8:07

Now that we know how to find the GCF, we are going to learn techniques for factoring out the GCF.

To do this requires the _____ property in_____ .

6. **Example:** Factor: $7z^2 - 14z$

   Note: Listen carefully to Step 1 regarding the factoring out of a negative with the GCF.

Write the steps in words	Show the steps with math
Step 1	
Step 2	
Step 3	
Step 4	

   Final answer: $7z^2 - 14z =$ _____

Steps to Factor a Polynomial Using the GCF

Step 1: Identify the _____ of the terms that make up the polynomial.

Step 2: Rewrite each term as the _____ of the _____ and the remaining _____. *Note: This is step is something you will eventually do in your head.*

Step 3: Use the _____ _____ "in reverse" to factor out the GCF.

Step 4: Check Use the Distributive Property.

Course: Name:
Instructor: Section:

7. **Example:** Factor the trinomial $36a^6 + 45a^4 - 18a^2$ by factoring out the GCF.

Final answer: $36a^6 + 45a^4 - 18a^2 = $ _____

Course: Name:
Instructor: Section:

Section 6.1 – Objective 2: Factor Out the Greatest Common Factor in Polynomials
Part II – Text Examples 9 and 10
Video Length – 4:56

8. **Example:** Factor the trinomial $-3x^6 + 9x^4 - 18x$ by factoring out the GCF.

   Note: When the highest degreed term has a negative coefficient, factor the negative out as part of the GCF.

   Final answer: $-3x^6 + 9x^4 - 18x = $ _____

9. **Example:** Factor out the greatest common binomial factor: $6(3x+y) - z(3x+y)$

   Final answer: $6(3x+y) - z(3x+y) = $ _____

Course: Name:
Instructor: Section:

Section 6.1 – Objective 3: Factor Polynomials by Grouping
Part I – Text Example 11
Video Length – 5:49

Sometimes a common factor is not going to occur in each of the terms in a polynomial, but a common factor will occur within a subset or a smaller grouping of the terms. When that happens we use the technique called **factoring by grouping**.

10. **Example:** Factor by grouping: $5x - 5y + ax - ay$

Write the steps in words	Show the steps with math
Step 1	
Step 2	
Step 3	
Step 4	

Final answer: $5x - 5y + ax - ay = $ _____

Steps to Factor a Polynomial by Grouping

Step 1: _____ the terms with _____ _____.

Step 2: In each grouping, _____ _____ the greatest common factor.

Step 3: If the remaining factor in each grouping is the same, _____ _____ _____.

Step 4: _____ _____ _____ by finding the product of factors.

Course: Name:
Instructor: Section:

Section 6.1 – Objective 3: Factor Polynomials by Grouping
Part II – Text Examples 12 and 13
Video Length – 11:08

We are going to consolidate the different factoring techniques that we have. Right now we have two factoring techniques. First and foremost is finding the Greatest Common Factor; factoring it out if it exists. Method two is factoring by grouping.

Whenever you are factoring, _____!

11. Example: Factor by grouping: $4x^3 - 8x^2 + 6x - 12$

Final answer: $4x^3 - 8x^2 + 6x - 12 =$ _____

Note: After completing this problem a student asks, "Do we have to take out the two from all of them in the beginning...will we still get the same answer?" Let's see. For the same example, first use factor by grouping. Then factor out the GCF.

12. Example: Factor by grouping: $4x^3 - 8x^2 + 6x - 12$

Final answer: $4x^3 - 8x^2 + 6x - 12 =$ _____

Copyright © 2014 Pearson Education, Inc.

Course: Name:
Instructor: Section:

Section 6.2 Video Guide
Factoring Trinomials of the Form $x^2 + bx + c$

Objectives:
1. Factor Trinomials of the Form $x^2 + bx + c$.
2. Factor out the Greatest Common Factor, Then Factor $x^2 + bx + c$.

Section 6.2 – Introduction
Video Length – 1:47

This section is dedicated to factoring trinomials, where the trinomial is of the form _____ .

In other words, the coefficient on the squared term is always going to equal ___ .

Definition
A _____ _____ is a polynomial of the form _____ , $a \neq 0$ where a represents the coefficient of the squared (second degree) term, b represents the coefficient of the linear (first degree) term and c represents the constant.

Examples of quadratic trinomials:

Definition
When the trinomial is written in standard form (or descending order of degree), the coefficient of the squared term is called the _____ _____ . Our goal for this section is going to be to factor quadratic trinomials of the form $x^2 + bx + c$.

Course:
Instructor:
Name:
Section:

Section 6.2 – Objective 1: Factor Trinomials of the Form $x^2 + bx + c$
Part I – Text Examples 1 and 2
Video Length – 10:23

Remember, whenever we do something in math, we always want to have the ability to undo it. For example, consider

$$(x+3)(x+5) = \underline{\hspace{3cm}}$$

1. **Example:** Factor: $x^2 + 8x + 12$

   Note: Pay special attention to the comment (after Step 3) about the amount of work needed.

Write the steps in words	Show the steps with math
Step 1	
Step 2	
Step 3	

   Final Answer: $x^2 + 8x + 12 = \underline{\hspace{4cm}}$

   Note: When the constant is positive, and the coefficient on the linear term is positive, both factors **must** be positive. Nice!

Steps to Factor a Trinomial of the Form $x^2 + bx + c$

Step 1: Find the pair of _____ whose _____ is ___ and whose _____ is _____. That is, determine m and n such that _____ = ___ and _____ = ___.

Step 2: Write _____ = _____ .

Step 3: Check your work by multiplying the binomials.

In a nutshell, you want to find factors of the constant, whose sum is the middle term.

Course:
Instructor:

Name:
Section:

2. **Example:** Factor: $n^2 - 10n + 21$

Final Answer: $n^2 - 10n + 21 = $ _____

Course:
Instructor:

Name:
Section:

Section 6.2 – Objective 1: Factor Trinomials of the Form $x^2 + bx + c$
Part II – Text Examples 3 and 4
Video Length – 5:05

3. **Example:** Factor: $x^2 + x - 42$

   Final Answer: $x^2 + x - 42 =$ _____

4. **Example:** Factor: $x^2 - 3x - 28$

   Final Answer: $x^2 - 3x - 28 =$ _____

Trinomials of the Form $x^2 + bx + c$
The following table summarizes the four forms for factoring a quadratic trinomial in the form $x^2 + bx + c$.

Form	Signs of *m* and *n*	Example
$x^2 + bx + c$, where *b* and *c* are both _____ .	*m* and *n* are both _____ .	
$x^2 + bx + c$, *b* is _____ and *c* is _____ .	*m* and *n* are both _____ .	
$x^2 + bx + c$, *b* is _____ and *c* is _____ .	*m* and *n* are _____ in sign and the factor with the larger absolute value is _____ .	
$x^2 + bx + c$, where *b* and *c* are both _____ .	*m* and *n* are _____ in sign and the factor with the larger absolute value is _____ .	

Course: Name:
Instructor: Section:

Section 6.2 – Objective 1: Factor Trinomials of the Form $x^2 + bx + c$
Part III – Text Examples 5, 6, and 7
Video Length – 4:34

Definition
A polynomial that cannot be written as the product of two other polynomials (other than 1 or −1) is said to be a
_____ _____ .

5. **Example:** So how do we identify a prime polynomial? Consider the polynomial $z^2 + 3z + 14$.

   Final Answer: _____
   Note: Write your final answer as a complete sentence.

6. **Example:** Factor: $x^2 + 5xy - 24y^2$

   Final Answer: $x^2 + 5xy - 24y^2 =$ _____

Course: Name:
Instructor: Section:

Section 6.2 – Objective 2: Factor Out the GCF, then Factor $x^2 + bx + c$
Video Length – 4:15

Remember, whenever you are factoring the first thing that you always want to do is look for the Greatest Common Factor.

7. **Example:** Factor: $2x^2 - 32x + 96$

   Final Answer: $2x^2 - 32x + 96 = $ _____

8. **Example:** Factor: $-x^2 - 12x - 36$
   Note: Be careful! You don't want to make any sign errors.

   Final Answer: $-x^2 - 12x - 36 = $ _____

Course: Name:
Instructor: Section:

Section 6.3 Video Guide
Factoring Trinomials of the Form ax^2+bx+c, $a \neq 1$

Objectives:
1. Factor ax^2+bx+c, $a \neq 1$, Using Grouping
2. Factor ax^2+bx+c, $a \neq 1$, Using Trial and Error

Section 6.3 – Introduction
Video Length – 0:34

When factoring trinomials of the form ax^2+bx+c where a is not equal to 1, there are two methods that can be used:

1. _____ and _____ ; 2. _____ by _____

Course: Name:
Instructor: Section:

Section 6.3 – Objective 1: Factor ax^2+bx+c, $a \neq 1$, Using Grouping
Video Length – 15:40

The second method that can be used to factor trinomials of the form ax^2+bx+c is **factoring by grouping**. This method has an advantage because its approach to factoring is algorithmic. This means that you just need to follow a bunch of steps to ultimately end up with the factored form.

1. **Example:** Factor $3x^2+17x+10$ using the Grouping Method.

Write the steps in words	Show the steps with math
Step 1	
Step 2	
Step 3	
Step 4	
Step 5	

Final Answer: $3x^2+17x+10 =$ _____

Course: Name:
Instructor: Section:

Factoring $ax^2 + bx + c$, $a \neq 1$, **by Grouping:** a, b, **and** c **Have No Common Factors**

Step 1: Find the value of _____ .

Step 2: Find the pair of integers, m and n, whose _____ is _____ and whose is _____ is _____.

Step 3: Write $ax^2 + bx + c =$ ____ + ____ + ____ + ____ .

Step 4: Factor the expression in Step 3 by grouping.

Step 5: _____ by multiplying the factors.

Note: An extremely, EXTREMELY, helpful suggestion in given in Step 3. Pay close attention to it.

2. **Example:** Factor by grouping: $6x^2 - x - 2$

    Final Answer: $6x^2 - x - 2 =$ _____

3. **Example:** Factor by grouping: $8x^2 + 10x - 3$

    Final Answer: $8x^2 + 10x - 3 =$ _____

Course:
Instructor:
Name:
Section:

Section 6.3 – Objective 2: Factor $ax^2 + bx + c$, $a \neq 1$, Using Trial and Error
Part I – Text Examples 5 and 6
Video Length – 9:43

4. **Example:** Factor: $2x^2 + 7x + 3$

Write the steps in words	Show the steps with math
Step 1	
Step 2	
Step 3	

Final Answer: $2x^2 + 7x + 3 = $ _____

Factoring $ax^2 + bx + c$, $a \neq 1$, Using Trial and Error: a, b, and c Have No Common Factors

Step 1: List the possibilities for the _____ _____ of each binomial whose product is

_____.

$$(\_x + \quad)(\_x + \quad) = ax^2 + bx + c$$

Step 2: List the possibilities for the _____ _____ of each binomial whose product is

_____.

$$(\_x + \square)(\_x + \square) = ax^2 + bx + c$$

Step 3: Write out all the combinations of factors found in Steps 1 and 2. Multiply the binomials out until a product is found that equals the trinomial.

Course:
Instructor:

Name:
Section:

5. Example: Factor by trial and error: $5x^2 - 17x + 6$

Final Answer: $5x^2 - 17x + 6 = $ _____

Course:
Instructor:
Name:
Section:

Section 6.3 – Objective 2: Factor ax^2+bx+c, $a \neq 1$, Using Trial and Error
Part II – Text Examples 7, 8, 9, and 10
Video Length – 10:53

Note: Listen carefully to what is said before the start of Step 3. Suggestions are made that can save you a lot of time when factoring by trial and error.

6. **Example:** Factor by trial and error: $6x^2 - x - 2$

   Final Answer: $6x^2 - x - 2 =$ _____

7. **Example:** Factor by trial and error: $24y^2 - 14y - 3$

   Final Answer: $24y^2 - 14y - 3 =$ _____

Course: Name:
Instructor: Section:

Section 6.4 Video Guide
Factoring Special Products

Objectives:
1. Factor Perfect Square Trinomials
2. Factor the Difference of Two Squares
3. Factor the Sum or Difference of Two Cubes

Section 6.4 – Objective 1: Factor Perfect Square Trinomials
Video Length – 11:41

In this section we will focus on factoring "special products." The first special product we will focus on is a **perfect square trinomial**.

Perfect Square Trinomials

$$a^2 + 2ab + b^2 = ()^2$$

$$a^2 - 2ab + b^2 = ()^2$$

In order for a polynomial to be a perfect square trinomial, two conditions must be satisfied:

1. The _____ and _____ terms must be _____ _____ .

   Examples of perfect squares are,

2. The "_____ term" must equal _____ or _____ times the product of the expressions being squared in the first and last term.

   For example,

1. **Example:** Factor the following:

   (a) $x^2 + 10x + 25$ (a) **Final Answer:** $x^2 + 10x + 25 = $ _____

   (b) $9x^2 - 30xz + 25z^2$ (b) **Final Answer:** $9x^2 - 30xz + 25z^2 = $ _____

Course: Name:
Instructor: Section:

Section 6.4 – Objective 2: Factor the Difference of Two Squares
Video Length – 3:39

Difference of Two Squares
$$a^2 - b^2 = ()()$$

2. **Example:** Factor the following:

   (a) $3x^2 - 27$ (a) **Final Answer:** $3x^2 - 27 = $ _____

   (b) $4x^2 - 25y^4$ (b) **Final Answer:** $4x^2 - 25y^4 = $ _____

Course: Name:
Instructor: Section:

Section 6.4 – Objective 3: Factor the Sum or Difference of Two Cubes
Video Length – 3:31

The Sum of Two Cubes
$$A^3 + B^3 = ()()$$

3. **Example:** Factor: $27x^3 + 125$

   Final Answer: $27x^3 + 125 = $ _____

The Difference of Two Cubes
$$A^3 - B^3 = ()()$$

4. **Example:** Factor: $27x^3 - 64y^6$

   Final Answer: $27x^3 - 64y^6 = $ _____

Course: Name:
Instructor: Section:

Section 6.5 Video Guide
Summary of Factoring Techniques

Objectives:
1. Factor Polynomial Completely

Section 6.5 – Introduction
Video Length – 4:59

Steps for Factoring

Step 1: Factor out the _____ _____ _____ (_____), if any exists.

Step 2: Count the _____ _____ _____ .

Step 3: (a) 2 terms

- Is it the difference of two squares? If so,

    _____ = _____

- Is it the difference of two cubes? If so,

    _____ = _____

- Is it the sum of two cubes? If so,

    _____ = _____

(b) 3 terms

- Is it a perfect square trinomial? If so,

    _____ = _____ or _____ = _____

- Is the coefficient of the square term 1? If so,

    _____ = _____ where ____ = ___ and ____ = ___

- Is the coefficient of the square term $\neq 1$? If so,

    a. Use _____ _____ _____

    b. Use _____ _____ _____

(c) 4 terms

- Use _____ _____ _____

Step 4: _____ _____ _____ by multiplying out the factored form.

Course: Name:
Instructor: Section:

Section 6.5 – Objective 1: Factor Polynomials Completely
Part I – Text Example 1
Video Length – 4:20

1. **Example:** Factor completely: $6x^2 - 6x - 36$

   Final Answer: $6x^2 - 6x - 36 = $ _____

Course:
Instructor:

Name:
Section:

Section 6.5 – Objective 1: Factor Polynomials Completely
Part II – Text Example 2
Video Length – 7:15

2. **Example:** Factor completely: $8x^2 - 18x - 35$

Final Answer: $8x^2 - 18x - 35 =$ _____

Course:
Instructor:

Name:
Section:

Section 6.5 – Objective 1: Factor Polynomials Completely
Part III – Text Example 4
Video Length – 4:44

3. **Example:** Factor completely: $80w^3 - 10$

Final Answer: $80w^3 - 10 =$ _____

Note: It is mentioned that the trinomial after factoring the difference of two cubes (or sum of two cubes) is always prime. In other words, it will not factor over the reals. This is true only if the terms have no common factor. This is another reason why factoring out the GCF in the first step is highly important.

Course:
Instructor:

Name:
Section:

Section 6.5 – Objective 1: Factor Polynomials Completely
Part IV – Text Example 5
Video Length – 3:39

4. **Example:** Factor completely: $1-16x^4$

Final Answer: $1-16x^4 = $ _____

WATCH OUT! _____

_____

Course: Name:
Instructor: Section:

Section 6.6 Video Guide
Solving Polynomial Equations by Factoring

Objectives:
1. Solve Quadratic Equations Using the Zero-Product Property
2. Solve Polynomial Equations of Degree Three or Higher Using the Zero-Product Property

---

Section 6.6 – Introduction
Video Length – 1:17

We will now utilize our factoring skills to help solve a new type of equation. This new type of equation is called a **polynomial equation**.

Definition
A _____ _____ is any equation that contains a polynomial expression. The _____ of a polynomial equation is the degree of the polynomial expression in the equation.

Examples of polynomial equations:

Course: Name:
Instructor: Section:

Section 6.6 – Objective 1: Solve Quadratic Equations Using the Zero-Product Property
Part I
Video Length – 3:13

The Zero-Product Property
If the product of two factors is zero, then at least one of the factors is _____ . That is, if

___ = ___ , then ___ = ___ or ___ = ___ or both ___ and ___ are ___ .

1. **Example:** Solve: $(x-5)(x+4)=0$

Final Answer: _____

Course: Name:
Instructor: Section:

Section 6.6 – Objective 1: Solve Quadratic Equations Using the Zero-Product Property
Part II – Text Example 1
Video Length – 5:06

Definition
A _____ _____ is an equation that can be written in the form (called

_____ _____)

where a, b, and c are real numbers and $a \neq 0$.

Quadratic equations are also referred to as _____ _____ equations because the

polynomial expression in the equation is a polynomial of degree _____ .

2. **Example:** Solve: $x^2 + 3x = -2$

Write the steps in words	Show the steps with math
Step 1	
Step 2	
Step 3	
Step 4	
Step 5	

Final Answer: _____

Course: Name:
Instructor: Section:

Section 6.6 – Objective 1: Solve Quadratic Equations Using the Zero-Product Property
Part III – Text Example 4
Video Length – 3:58

3. **Example:** Solve: $2m^2 - 3m - 7 = 2m + 5$

Final Answer: _____

Course: Name:
Instructor: Section:

Section 6.6 – Objective 1: Solve Quadratic Equations Using the Zero-Product Property
Part IV – Text Example 5
Video Length – 4:07

4. **Example:** Solve: $(x+3)(3x+5) = 7$

   PLEASE PLEASE PLEASE avoid the urge to set each factor equal to 7.

   Final Answer: _____

Course: Name:
Instructor: Section:

Section 6.6 – Objective 1: Solve Quadratic Equations Using the Zero-Product Property
Part V – Text Example 6
Video Length – 4:08

Based on our experience, we might conjecture that when solving a quadratic equation we will get two solutions. Is that always the case? Let's solve the following example to see what happens.

5. **Example:** Solve: $9a^2 + 25 = 30a$

   Final Answer: _____

Course:
Instructor:

Name:
Section:

Section 6.6 – Objective 2: Solve Polynomial Equations of Degree Three or Higher Using the Zero-Product Property

Video Length – 8:52

6. **Example:** Solve: $3x^3 + x^2 = 14x$

Write the steps in words	Show the steps with math
Step 1	
Step 2	

Final Answer: _____

7. **Example:** Solve: $3x^3 + 15x^2 - 27x - 135 = 0$

Final Answer: _____

Course: Name:
Instructor: Section:

Section 6.7 Video Guide
Modeling and Solving Problems with Quadratic Equations

Objectives:
1. Model and Solve Problems Involving Quadratic Equations
2. Model and Solve Problems Using the Pythagorean Theorem

Section 6.7 – Objective 1: Model and Solve Problems Involving Quadratic Equations
Video Length – 7:08

1. **Example**: The area of a rectangle is 84 square inches. Determine the length and width if the length is 2 inches less than twice the width.

Write the steps in words	Show the steps with math
Step 1	
Step 2	
Step 3	
Step 4	
Step 5 Note: He checks the work mentally. But you should show your work in the space provided to the right.	
Step 6	

220 Copyright © 2014 Pearson Education, Inc.

Course:
Instructor:
Name:
Section:

Section 6.7 – Objective 2: Model and Solve Problems Using the Pythagorean Theorem
Video Length – 6:22

Definition
A _____ _____ is one that contains a right angle, that is, an angle of _____ .

The side of the triangle opposite the 90° angle is called the _____ ; the remaining two

sides are called the _____ .

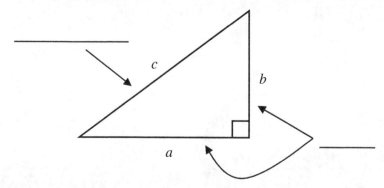

Pythagorean Theorem
In a right triangle, the square of the length of the hypotenuse is equal to the sum of the squares of the lengths of the legs.

___ = ___ + ___ or _____ + _____ = _____

2. **Example:** Find the length of each leg of the right triangle.

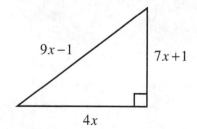

Final Answer: _____
Note: Write your final answer as a complete sentence.

Course: Name:
Instructor: Section:

Section 7.1 Video Guide
Simplifying Rational Expressions

Objectives:
1. Evaluate a Rational Expression
2. Determine Values for Which a Rational Expression Is Undefined
3. Simplify Rational Expressions

Section 7.1 – Objective 1: Evaluate a Rational Expression
Video Length – 5:04

We are now going work with **rational expressions**.

Note: It is HIGHLY RECOMMENDED that you review how to simplify, add, subtract, multiply, and divide fractions. Everything you know about fractions will be extended to rational expressions.

Definition
A _____ _____ is the _____ of two polynomials. That is, a rational expression is written in the form $\frac{p}{q}$, where p and q are polynomials and $q \neq 0$.

Examples of rational expressions:

Definition
To _____ a rational expression, replace the variable with its assigned numerical value and perform the arithmetic.

1. **Example:** Evaluate $\dfrac{x+2}{3x-5}$ for $x=4$.

    Final answer: _____

2. **Example:** Evaluate $\dfrac{6x^2 - 3xy + 10y^2}{x+y}$ for $x=-1$ and $y=3$.

    Final answer: _____

Course: Name:
Instructor: Section:

Section 7.1 – Objective 2: Determine Values for Which a Rational Expression Is Undefined
Video Length – 5:15

Because a rational expression is undefined for those values of the variable(s) that make the denominator zero , **we find the values for which a rational expression is _____ by setting the _____ equal to _____ and solving for the variable**.

3. **Example:** Find the value for which $\dfrac{x+2}{3x-5}$ is undefined.

   Final answer: _____.
   Note: Write your final answer as a complete sentence.

4. **Example:** Find the value(s) of x for which $\dfrac{2x+3}{x^2-2x-24}$ is undefined.

   Final answer: _____.
   Note: Write your final answer as a complete sentence.

Course: Name:
Instructor: Section:

Section 7.1 – Objective 3: Simplify Rational Expressions
Part I – Text Examples 6, 7, and 8
Video Length – 6:09

We will now simplify rational expressions. Consider the rational number

$$\frac{10}{35}$$

We will use the exact same logic to simplify rational expressions.

Simplifying Rational Expressions
If p, q, and r are polynomials, then

$$\frac{}{} = \frac{}{} \text{ if } q \neq 0 \text{ and } r \neq 0.$$

5. **Example:** Simplify: $\dfrac{5x+10}{x+2}$, $x \neq -2$.

Write the steps in words	Show the steps with math
Step 1	
Step 2	

Final answer: $\dfrac{5x+10}{x+2} = $ _____

6. **Example:** Simplify: $\dfrac{x^3 - 3x^2 - 40x}{x^2 + 10x + 25}$

Final answer: $\dfrac{x^3 - 3x^2 - 40x}{x^2 + 10x + 25} = $ _____

Course:
Instructor:
Name:
Section:

Section 7.1 – Objective 3: Simplify Rational Expressions
Part II – Text Example 9
Video Length – 4:44

Note: In general, $a-b$ CANNOT simply be rewritten as $b-a$. So pay VERY CLOSE ATTENTION to how he rewrites the difference of two terms.

Consider the following expressions

$$3 + 2x$$

$$2 - z$$

$$\frac{z-2}{2-z}$$

7. **Example:** Simplify: $\dfrac{9-x^2}{2x^2 - 7x + 3}$

Final answer: $\dfrac{9-x^2}{2x^2 - 7x + 3} = $ _____

Course: Name:
Instructor: Section:

Section 7.2 Video Guide
Multiplying and Dividing Rational Expressions

Objectives:
1. Multiply Rational Expressions
2. Divide Rational Expressions

Section 7.2 – Objective 1: Multiply Rational Expressions
Video Length – 7:44

We will now multiply rational expressions. Multiplying rational expressions follows the exact same logic that we use to multiply rational numbers. For example,

$$\frac{2}{3} \cdot \frac{27}{8}$$

1. **Example:** Multiply: $\dfrac{x+2}{7} \cdot \dfrac{7x+21}{x^2-4}$

Write the steps in words	Show the steps with math
Step 1	
Step 2	
Step 3	
Note: PLEASE pay attention to his "warning."	

Final answer: $\dfrac{x+2}{7} \cdot \dfrac{7x+21}{x^2-4} = $ _____

Copyright © 2014 Pearson Education, Inc.   227

Course: Name:
Instructor: Section:

2. **Example:** Multiply: $\dfrac{7x+7}{x+4} \cdot \dfrac{x^2-x-20}{7x^2-42x-49}$

Final answer: $\dfrac{7x+7}{x+4} \cdot \dfrac{x^2-x-20}{7x^2-42x-49} =$ _____

Course: Name:
Instructor: Section:

Section 7.2 – Objective 2: Divide Rational Expressions
Video Length – 9:47

Determine the following:

$$\frac{3}{4} \div \frac{20}{9}$$

3. **Example:** Divide: $\dfrac{x+2}{5x} \div \dfrac{2x+4}{15x^2}$

Final answer: $\dfrac{x+2}{5x} \div \dfrac{2x+4}{15x^2} =$ _____

Steps to Divide Rational Expressions

Step 1: Multiply the dividend by the _____ of the divisor. That is,

$$— \div — = — \cdot —$$

Step 2: _____ each polynomial in the numerator and denominator.

Step 3: _____ .

Step 4: _____ ____ _____ _____ in the numerator and denominator. Leave the remaining factors in factored form.

Note: Heads up! The next problem looks crazy. But it ONLY looks crazy. It is not as bad as you may think it is.

Course: Name:
Instructor: Section:

4. **Example:** Divide: $\dfrac{\dfrac{a^2+2a-8}{2a^2+a-3}}{\dfrac{a^2+4a}{2a^2-a-6}}$

Final answer: $\dfrac{\dfrac{a^2+2a-8}{2a^2+a-3}}{\dfrac{a^2+4a}{2a^2-a-6}} = $ _____

Course: Name:
Instructor: Section:

Section 7.3 Video Guide
Adding and Subtracting Rational Expressions with a Common Denominator

Objectives:
1. Add Rational Expressions with a Common Denominator
2. Subtract Rational Expressions with a Common Denominator
3. Add or Subtract Rational Expressions with Opposite Denominators

Section 7.3 – Objective 1: Add Rational Expressions with a Common Denominator
Video Length – 4:58

Determine the following:

$$\frac{5}{9} + \frac{7}{9}$$

The approach to adding rational expressions is exactly the same.

1. **Example:** Add: $\dfrac{4}{x+3} + \dfrac{6}{x+3}$

Write the steps in words	Show the steps with math
Step 1	
Step 2	

Final answer: $\dfrac{4}{x+3} + \dfrac{6}{x+3} = $ _____

Course: Name:
Instructor: Section:

2. **Example:** Add: $\dfrac{3a^2+a}{a^2-9}+\dfrac{2a-2a^2}{a^2-9}$

Final answer: $\dfrac{3a^2+a}{a^2-9}+\dfrac{2a-2a^2}{a^2-9} =$ _____

Course: Name:
Instructor: Section:

Section 7.3 – Objective 2: Subtract Rational Expressions with a Common Denominator
Video Length – 5:54

Determine the following:

$$\frac{8}{3} - \frac{2}{3}$$

We will now do the same with rational expressions.

3. **Example:** Subtract: $\dfrac{3k^2}{k^2 - k - 2} - \dfrac{6k}{k^2 - k - 2}$

Final answer: $\dfrac{3k^2}{k^2 - k - 2} - \dfrac{6k}{k^2 - k - 2} = $ _____

Note: Be really careful when subtracting a rational expression with two or more terms in the numerator. The following example illustrates why it is necessary to incorporate parentheses in your work.

4. **Example:** Subtract: $\dfrac{7x - 3}{x + 2} - \dfrac{3x - 11}{x + 2}$

Final answer: $\dfrac{7x - 3}{x + 2} - \dfrac{3x - 11}{x + 2} = $ _____

Copyright © 2014 Pearson Education, Inc.

Course: Name:
Instructor: Section:

Section 7.3 – Objective 3: Add or Subtract Rational Expressions with Opposite Denominators

Video Length – 5:06

5. **Example:** Add: $\dfrac{n^2 - 2n}{n-4} + \dfrac{8}{4-n}$

Final answer: $\dfrac{n^2 - 2n}{n-4} + \dfrac{8}{4-n} = $ _____

6. **Example:** Subtract: $\dfrac{p^2}{p^2 - 81} - \dfrac{9p}{81 - p^2}$

Final answer: $\dfrac{p^2}{p^2 - 81} - \dfrac{9p}{81 - p^2} = $ _____

Course: Name:
Instructor: Section:

Section 7.4 Video Guide
Finding the Least Common Denominator and Forming Equivalent Rational Expressions

Objectives:
1. Find the Least Common Denominator of Two or More Rational Expressions
2. Write a Rational Expression Equivalent to a Given Rational Expression
3. Use the LCD to Write Equivalent Rational Expressions

Section 7.4 – Objective 1: Find the Least Common Denominator of Two or More Rational Expressions
Part I – Text Examples 1, 2, and 3
Video Length – 10:29

1. **Example:** Find the LCD: $\frac{5}{12}$ and $\frac{8}{15}$.

Write the steps in words	Show the steps with math
Step 1	
Step 2	
Step 3	

   Final answer: LCD = _____

If the denominators of a sum or difference of rational expressions are not the same, the rational expressions must be written using a **least common denominator**.

Definition
The **least common denominator (LCD)** of two or more rational expressions is the _____ of _____ _____ that is a multiple of each denominator in the expressions.

Let's repeat what we just did with rational numbers, but this time using monomial denominators.

Course:
Instructor:

Name:
Section:

2. **Example:** Find the LCD: $\dfrac{1}{6a^2b}$ and $\dfrac{2}{9ab^3}$.

Final answer: LCD = _____

Course:
Instructor:

Name:
Section:

Section 7.4 – Objective 1: Find the Least Common Denominator of Two or More Rational Expressions
Part II – Text Examples 4, 5, and 6
Video Length – 5:25

3. **Example:** Find the LCD of $\dfrac{5}{x^2 - x - 6}$ and $\dfrac{9}{x^2 + 3x + 2}$.

   Final answer: LCD = _____

4. **Example:** Find the LCD of $\dfrac{4}{n^2 - 9}$ and $\dfrac{8}{21 - 7n}$.

   Final answer: LCD = _____

Course: Name:
Instructor: Section:

Section 7.4 – Objective 2: Write a Rational Expression Equivalent to a Given Rational Expression
Video Length – 13:22

5. **Example:** Write $\dfrac{8}{3}$ as an equivalent fraction with a denominator of 12.

   Final answer: $\dfrac{8}{3}$ = _____

6. **Example:** Write $\dfrac{7}{3x^2 + 2x}$ as an equivalent rational expression with a denominator of $12x^3 + 8x^2$.

Write the steps in words	Show the steps with math
Step 1	
Step 2	
Step 3	
Step 4 NOTE: It is not readily apparent why we would leave the denominator in factored form. However, it will definitely make our work much easier in later sections. Promise.	

   Final answer: $\dfrac{7}{3x^2 + 2x}$ = _____

Course: Name:
Instructor: Section:

7. **Example:** Write $\dfrac{2x+1}{x^2-3x-4}$ as an equivalent rational expression with $(x+1)^2(x-4)$ as the denominator.

Final answer: $\dfrac{2x+1}{x^2-3x-4} =$ _____

Course: Name:
Instructor: Section:

Section 7.4 – Objective 3: Use the LCD to Write Equivalent Rational Expressions
Video Length – 4:44

8. **Example:** Find the LCD of the rational expressions $\dfrac{6}{x^2+x}$ and $\dfrac{7}{x^2-x-2}$. Rewrite each expression.

Write the steps in words	Show the steps with math
Step 1	
Step 2	

Final answer: $\dfrac{6}{x^2+x} =$ _____ and $\dfrac{7}{x^2-x-2} =$ _____

Course: Name:
Instructor: Section:

Section 7.5 Video Guide
Adding and Subtracting Rational Expressions with Unlike Denominators

Objectives:
1. Add and Subtract Rational Expressions with Unlike Denominators

---

Section 7.5 – Objective 1: Add and Subtract Rational Expressions with Unlike Denominators
Part I – Text Examples 1 and 2
Video Length – 7:58

1. **Example:** Evaluate: $\dfrac{5}{12} + \dfrac{2}{15}$

Write the steps in words	Show the steps with math
Step 1	
Step 2	
Step 3	
Step 4	

Final answer: $\dfrac{5}{12} + \dfrac{2}{15} = $ _____

These are the steps that we use to add rational numbers. Guess what? They are the exact same steps that we are going to use to add rational expressions.

Course: Name:
Instructor: Section:

2. **Example:** Evaluate: $\dfrac{3}{4y^3} + \dfrac{7}{18y}$

Write the steps in words	Show the steps with math
Step 1	
Step 2	
Step 3	
Step 4	

Final answer: $\dfrac{3}{4y^3} + \dfrac{7}{18y} =$ _____

Course:
Instructor:
Name:
Section:

Section 7.5 – Objective 1: Add and Subtract Rational Expressions with Unlike Denominators
Part II – Text Examples 3 and 4
Video Length – 7:24

3. **Example:** Find the sum: $\dfrac{-2}{x+5} + \dfrac{3}{x+4}$

Final answer: $\dfrac{-2}{x+5} + \dfrac{3}{x+4} = $ _____

4. **Example:** Find the sum: $\dfrac{3}{x+2} + \dfrac{8-2x}{x^2-4}$

Final answer: $\dfrac{3}{x+2} + \dfrac{8-2x}{x^2-4} = $ _____

Course: Name:
Instructor: Section:

Section 7.5 – Objective 1: Add and Subtract Rational Expressions with Unlike Denominators
Part III – Text Example 5
Video Length – 5:42

5. **Example:** Subtract: $\dfrac{3}{x^2+7x+10} - \dfrac{4}{x^2+6x+5}$

Write the steps in words	Show the steps with math
Step 1	
Step 2	
Step 3	
Step 4	

Final answer: $\dfrac{3}{x^2+7x+10} - \dfrac{4}{x^2+6x+5} =$ _____

Course: Name:
Instructor: Section:

Section 7.6 Video Guide
Complex Rational Expressions

Objectives:
1. Simplify a Complex Rational Expression by Simplifying the Numerator and Denominator Separately (Method I)
2. Simplify a Complex Rational Expression by Using the Least Common Denominator (Method II)

Section 7.6 – Objective 1: Simplify a Complex Rational Expression by Simplifying the Numerator and Denominator Separately (Method I)
Part I – Text Example 1
Video Length – 1:14

1. **Example:** Divide: $\dfrac{\dfrac{4x}{x-3}}{\dfrac{2x}{x-3}}$

Final answer: $\dfrac{\dfrac{4x}{x-3}}{\dfrac{2x}{x-3}} = $ _____

Course: Name:
Instructor: Section:

Section 7.6 – Objective 1: Simplify a Complex Rational Expression by Simplifying the Numerator and Denominator Separately (Method I)
Part II – Text Example 2
Video Length – 9:11

Definition
A _____ _____ _____ is a fraction in which the numerator and/or the denominator contains the sum or difference of two or more rational expressions.

2. **Example:** Simplify: $\dfrac{\frac{1}{2}+3}{\frac{3}{8}+1}$

Final answer: $\dfrac{\frac{1}{2}+3}{\frac{3}{8}+1} =$ _____

Course: Name:
Instructor: Section:

3. **Example:** Simplify: $\dfrac{\dfrac{4}{c}}{\dfrac{7}{c}+\dfrac{2}{c^2}}$

Write the steps in words	Show the steps with math
Step 1	
Step 2	
Step 3	
Step 4	

Final answer: $\dfrac{\dfrac{4}{c}}{\dfrac{7}{c}+\dfrac{2}{c^2}} =$ _____

Course: Name:
Instructor: Section:

Section 7.6 – Objective 1: Simplify a Complex Rational Expression by Simplifying the Numerator and Denominator (Method I)
Part III – Text Examples 3 and 4
Video Length – 10:32

4. **Example:** Simplify: $\dfrac{\dfrac{2}{a^2}+\dfrac{3}{a}}{\dfrac{1}{a^3}-\dfrac{4}{a}}$

Final answer: $\dfrac{\dfrac{2}{a^2}+\dfrac{3}{a}}{\dfrac{1}{a^3}-\dfrac{4}{a}} =$ _____

Course:
Instructor:

Name:
Section:

5. **Example:** Simplify: $\dfrac{\dfrac{1}{x+3}+1}{x-\dfrac{4}{x+3}}$

Final answer: $\dfrac{\dfrac{1}{x+3}+1}{x-\dfrac{4}{x+3}} = $ _____

Course: Name:
Instructor: Section:

Section 7.6 – Objective 2: Simplify a Complex Rational Expression Using the Least Common Denominator (Method II)
Part I – Text Example 5
Video Length – 12:01

Note: Remember the following example? It was simplified using Method I. Now we will show you another approach for simplifying complex rational expressions.

6. **Example:** Simplify: $\dfrac{\frac{1}{2}+3}{\frac{3}{8}+1}$

Write the steps in words	Show the steps with math
Step 1	
Step 2	

Final answer: $\dfrac{\frac{1}{2}+3}{\frac{3}{8}+1} = $ _____

Course: Name:
Instructor: Section:

7. **Example:** Simplify: $\dfrac{\dfrac{w}{8}-\dfrac{1}{2}}{\dfrac{3}{8w}+\dfrac{3}{4}}$

Final answer: $\dfrac{\dfrac{w}{8}-\dfrac{1}{2}}{\dfrac{3}{8w}+\dfrac{3}{4}} =$ _____

Course: Name:
Instructor: Section:

Section 7.6 – Objective 2: Simplify a Complex Rational Expression Using the Least Common Denominator (Method II)
Part II – Text Example 6
Video Length – 3:03

8. **Example:** Simplify: $\dfrac{\dfrac{1}{x+3}+1}{x-\dfrac{4}{x+3}}$

Final answer: $\dfrac{\dfrac{1}{x+3}+1}{x-\dfrac{4}{x+3}} = $ _____

Course: Name:
Instructor: Section:

Section 7.7 Video Guide
Rational Equations

Objectives:
1. Solve Equations Containing Rational Expressions
2. Solve for a Variable in a Rational Equation

Section 7.7 – Objective 1: Solve Equations Containing Rational Expressions
Part I – Text Examples 1 and 2
Video Length – 17:25

Up to this point, we learned how to solve linear equations. We also looked at solving quadratic equations and polynomial equations of degree three or higher by factoring. Now what we're going to do is solve **rational equations**.

Definition
A _____ _____ is an equation that contains a rational expression.

Note: In order to solve a rational equation, we will adopt the same method that was used to solve linear equations with fractional coefficients. For review, solve the following LINEAR equation.

1. **Example:** Solve: $\dfrac{x}{4} - \dfrac{2}{3}x = \dfrac{5}{6}$

   Final answer: _____

These are the exact steps that we are going to use to solve rational equations.

Note: There will be one major difference when solving rational equations. You will need to consider the possibility of division by zero. This will be illustrated in the following examples.

Course: Name:
Instructor: Section:

Note: Pay attention to what he says about division by zero

2. **Example:** Solve: $\dfrac{1}{3x} + \dfrac{5}{6} = \dfrac{2}{x}$

Write the steps in words	Show the steps with math
Step 1	
Step 2	
Step 3	
Step 4	
Step 5	

Final answer: _____

Course:
Instructor:

Name:
Section:

3. **Example:** Solve: $\dfrac{3}{y+5} = \dfrac{3}{3y-2}$

Final answer: _____

Note: He makes a comment about "cross multiplying." But this technique only works if the rational equation is written in a specific way. The steps he used above works for ANY rational equation and will ALWAYS work.

Course: Name:
Instructor: Section:

Section 7.7 – Objective 1: Solve Equations Containing Rational Expressions
Part II – Text Example 5
Video Length – 9:43

When solving rational equations you may come across something called an **extraneous solution**.

Definition
An _____ _____ is a solution that is obtained through the solving process that does not satisfy the _____ equation.

Note: Make sure you check for value(s) of the variable that result in division by zero.

4. **Example:** Solve: $\dfrac{2}{x+1} - \dfrac{5}{x-2} = \dfrac{x^2 - 6x - 13}{x^2 - x - 2}$

Final answer: _____

Course: Name:
Instructor: Section:

Section 7.7 – Objective 1: Solve Equations Containing Rational Expressions
Part III – Text Example 6
Video Length – 5:27

Note: Again, make sure you check for value(s) of the variable that result in division by zero.

5. **Example:** Solve: $\dfrac{-12}{x^2-9} = \dfrac{2}{x+3} - \dfrac{5}{x-3}$

Final answer: _____

Course: Name:
Instructor: Section:

Section 7.7 – Objective 2: Solve for a Variable in a Rational Equation
Video Length – 4:35

The expression "solve for the variable" means to get the variable by itself on one side of the equation with all other variables and constants, if any exist, on the other side. The steps that we follow when solving formulas for a certain variable are identical to those that we follow when solving rational equations.

6. **Example:** The formula

$$\frac{1}{f} = \frac{1}{p} + \frac{1}{q}$$

is used in telescope and camera construction, where f is the focal length of the lens, p is the distance between the object we wish to view and the lens, and q is the distance from the lens to the point of focus (such as your eye or film).

(a) Solve for the formula p.

(b) Suppose a camera has a focal length of 100 mm and the distance from the lens to the film is 105 mm. How far away is the object being photographed?

(a) **Final answer:** _____

(b) **Final answer:** _____
    Note: Write your answer as a complete sentence.

Course: Name:
Instructor: Section:

Section 7.8 Video Guide
Models Involving Rational Equations

Objectives:
1. Model and Solve Ratio and Proportion Problems
2. Model and Solve Problems with Similar Figures
3. Model and Solve Work Problems
4. Model and Solve Uniform Motion Problems

Section 7.8 – Objective 1: Model and Solve Ratio and Proportion Problems
Video Length – 7:08

Definition
A _____ is the quotient of two numbers or two quantities. The ratio of two numbers a and b can be written as

_____ or _____ or _____

Definition
A _____ is an equation of the form $\dfrac{}{} = \dfrac{}{}$, $b, d \neq 0$, where a, b, c, and d are called the terms of the proportion.

$$\dfrac{1}{2} = \dfrac{5}{10}$$

1. **Example:** Solve the proportion: 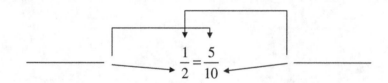 $\dfrac{n}{5} = \dfrac{12}{4}$

Final answer: _____

Course: Name:
Instructor: Section:

Another method that is frequently used to solve proportion problems is the method of cross multiplication.

Means-Extremes Theorem

If $\dfrac{a}{b} = \dfrac{c}{d}$, where $b \neq 0$ and $d \neq 0$, then ___ · ___ = ___ · ___ .

Definition
The Means-Extremes Theorem is also called the method of _____ _____ and is used to solve proportion problems.

Cross-Multiplication Property

If $\dfrac{a}{b} = \dfrac{c}{d}$, $b, d \neq 0$, then ___ · ___ = ___ · ___ .

2. **Example:** Solve the proportion:

$$\dfrac{15}{4} = \dfrac{n}{6}$$

Final answer: _____

3. **Example:** On a recent business trip to Japan, Yolanda purchased a jacket for 14,000 Yen (¥). At the time, one U.S. dollar was equal to 88.875 Yen. To the nearest dollar, how much did the jacket cost?

Final answer: _____
Note: Write your answer as a complete sentence.

Course: Name:
Instructor: Section:

Section 7.8 – Objective 2: Model and Solve Problems with Similar Figures
Video Length – 4:13

Definition
Two figures are _____ if their _____ are the _____ and their

_____ _____ are _____ .

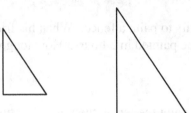

4. **Example:** Triangles A and B are similar. Find the length of side y.

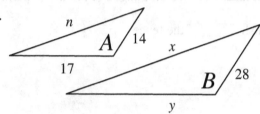

Final answer: _____

Course: Name:
Instructor: Section:

Section 7.8 – Objective 3: Model and Solve Work Problems
Video Length – 9:34

We are now going to solve work problems.

Note: Two "simplifying assumptions" are made to solve work problems. Make sure you know what they are and understand why they are made. Additionally, listen to how he checks to make sure the answer is reasonable.

5. **Example:** It takes John 4 hours to paint a fence. When his friend, Tom, helps him and the two work together, the fence can be painted in 3 hours. How long would it take Tom to paint the fence by himself?

   If it takes John 4 hours to paint the fence, then in _____ hour, he would finish _____ of the fence.

   If it takes both working together 3 hours to paint the fence, then in _____ hour, they would finish _____ of the fence.

   Final answer: _____
   Note: Write your answer as a complete sentence.

Course: Name:
Instructor: Section:

Section 7.8 – Objective 4: Model and Solve Uniform Motion Problems
Video Length – 7:26

We are now going to look at a uniform motion problem. Remember with uniform motion problems, we use the fact that ___ = ___ · ___ . In these problem we assume that the objects move at a

_____ speed.

6. **Example:** The speed of a boat in still water is 29 mph. If the boat travels 140 miles downstream in the same time that it takes to travel 92 miles upstream, find the speed of the stream.

Final answer: _____
Note: Write your answer as a complete sentence.

Course:
Instructor:

Name:
Section:

Course: Name:
Instructor: Section:

Getting Ready for Chapter 8 Video Guide
Interval Notation

Objectives:
1. Represent Inequalities Using the Real Number Line and Interval Notation

Getting Ready for Chapter 8 – Objective 1: Represent Inequalities Using the Real Number Line and Interval Notation
Part I – Text Examples 1 and 2
Video Length – 14:39

Definition
The notation $a < x < b$ means that x is between the two real numbers a and b. It is equivalent to the two inequalities $a < x$ and $x < b$. These expressions are in _____ _____.

Definition: Interval Notation
Let a and b represent two real number with $a < b$.

A _____ _____, denoted _____, consists of all real numbers x for which _____.

An _____ _____, denoted _____, consists of all real numbers x for which _____.

The _____ - _____ or _____ - _____ _____ are _____, consisting of all real number x for which _____, and _____, consisting of all real number x for which _____.

Course: Name:
Instructor: Section:

Interval	Interval Notation	Graph

1. **Example:** Write each inequality using interval notation and graph each interval.

   (a) $-2 < x \leq 0$

   Interval notation: _____

   Graph: ─────────▶

   (b) $x < 3$

   Interval notation: _____

   Graph: ─────────▶

   (c) $-1.5 < x < 3$

   Interval notation: _____

   Graph: ─────────▶

Course: Name:
Instructor: Section:

Getting Ready for Chapter 8 – Objective 1: Represent Inequalities Using the Real Number Line and Interval Notation
Part II – Text Examples 3 and 4
Video Length – 4:18

2. **Example:** Write each interval in inequality notation and graph each inequality.

   (a) $[-4, 2]$

       Inequality notation: _____

       Graph: ⎯⎯⎯⎯⎯⎯⎯⎯⎯⎯⎯⎯→

   (b)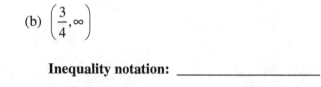

       Inequality notation: _____

       Graph: ⎯⎯⎯⎯⎯⎯⎯⎯⎯⎯⎯⎯→

   (c) $(0, 4]$

       Inequality notation: _____

       Graph: ⎯⎯⎯⎯⎯⎯⎯⎯⎯⎯⎯⎯→

Course:
Instructor:

Name:
Section:

Course: Name:
Instructor: Section:

Section 8.1 Video Guide
Graphs of Equations

Objectives:
1. Graph an Equation Using the Point-Plotting Method
2. Identify the Intercepts from the Graph of an Equation
3. Interpret Graphs

Section 8.1 – Objective 1: Graph an Equation Using the Point-Plotting Method
Video Length – 9:49

Remember, the graph of an equation is a graphical or geometrical way of representing the solution set to the equation in two variables.

One of the most elementary methods for graphing an equation is the **point-plotting method**. Values are chosen for one of the variables and the corresponding value of the remaining variable is determined by using the equation.

1. **Example**: Graph the equation $y = x + 3$ by plotting points.

2. **Example**: Graph the equation $y = x^2 + 3$ by plotting points.

Course: Name:
Instructor: Section:

Section 8.1 – Objective 2: Identify the Intercepts from the Graph of an Equation
Video Length – 4:59

Definition
The points, if any, at which the graph crosses or touches the coordinate axes are called the _____ .

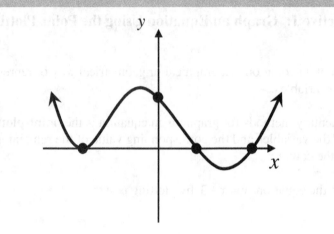

3. **Example:** Find the intercepts of the graph.

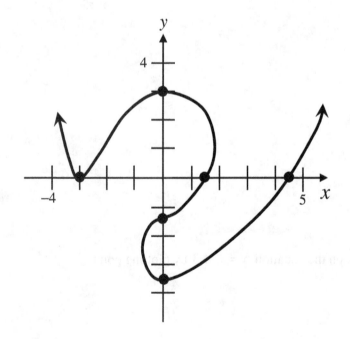

Note: Watch for the "Update" that pops up on the screen.

Course:
Instructor:
Name:
Section:

Section 8.1 – Objective 3: Interpret Graphs
Video Length – 3:41

Graphs play an important role in helping us to visualize relationships that exist between two variables or quantities.

4. **Example:** The graph below shows the yearly revenue R for selling cooking woks at a price p dollars. The vertical axis represents the revenue and the horizontal axis represents the price of each wok.

   (a) What is the revenue if the price of a work is $20?

   Final answer: _____
   Note: Remember, write your answer in a complete sentence.

   (b) What is the price of a wok when revenue is highest? What is the highest revenue?

   Final answer: _____
   Note: Remember, write your answer in a complete sentence.

   (c) Identify and interpret the intercepts.

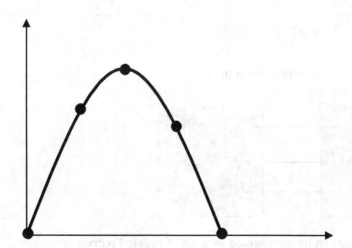

Note: Label the graph accordingly (e.g. label axes, plot and label tick marks, label ordered pairs, etc.)

Course: Name:
Instructor: Section:

Section 8.2 Video Guide
Relations

Objectives:
1. Understand Relations
2. Find the Domain and the Range of a Relation
3. Graph a Relation Defined by an Equation

Section 8.2 – Objective 1: Understand Relations
Video Length – 6:02

Definition
When the elements in one set are linked to elements in a second set of data, we have a

_____ .

Definition
If x and y are two elements in these sets and if a relation exists between x and y, then we say that x

_____ to y or that y _____ ____ x.

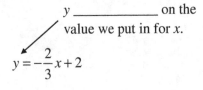

y _____ on the value we put in for x.

Below is a relation represented by a map.

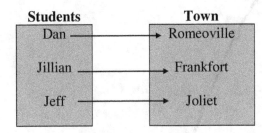

This same relation can be represented by a set of ordered pairs.

272

Course: Name:
Instructor: Section:

Section 8.2 – Objective 2: Find the Domain and the Range of a Relation
Part I – Text Examples 2 and 3
Video Length – 5:58

> **Definition**
> The _____ of a relation is the set of all _____ of the relation. The _____ is the set of all _____ of the relation.

1. **Example:** Find the domain and range of the relation.

   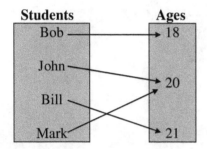

   Final answer: Domain: _____ Range: _____
   Note: After the range is determined, listen to what he says about listing elements in a set.

2. **Example:** Find the domain and range of the relation.

   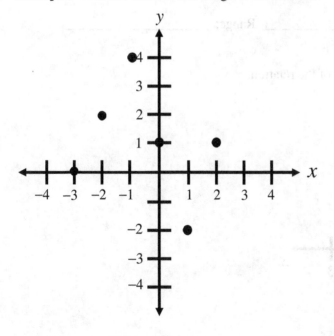

   Final answer: Domain: _____ Range: _____
   Note: Again, listen to what he says about listing elements in a set.

Copyright © 2014 Pearson Education, Inc.   273

Course:
Instructor:

Name:
Section:

Section 8.2 – Objective 2: Find the Domain and the Range of a Relation
Part II – Text Example 4
Video Length – 7:07

Relations can also be represented graphically.

3. **Example:** Find the domain and range of the relation.

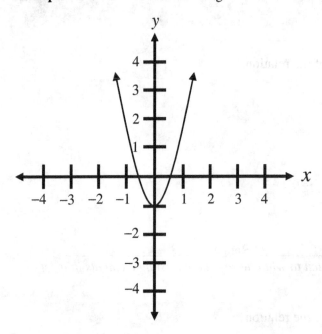

Final answer: Domain: _____ Range: _____

4. **Example:** Find the domain and range of the relation.

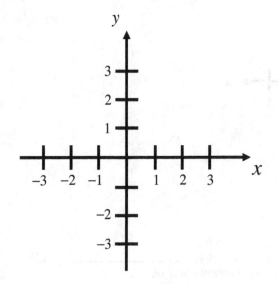

Final answer: Domain: _____ Range: _____

Course: Name:
Instructor: Section:

Section 8.2 – Objective 3: Graph a Relation Defined by an Equation
Video Length – 7:48

5. **Example:** Graph the relation $y = -\dfrac{2}{3}x + 2$. Determine the domain and range.

Final answer: Domain: _____ Range: _____

6. **Example:** Graph the relation $x - y^2 = 4$. Determine the domain and range.

Final answer: Domain: _____ Range: _____
Note: He asked for the domain of the "function", but he meant to say "relation."

Course:
Instructor:

Name:
Section:

Section 8.3 Video Guide
An Introduction to Functions

Objectives:
1. Determine Whether a Relation Expressed as a Map or Ordered Pairs Represents a Function
2. Determine Whether a Relation Expressed as an Equation Represents a Function
3. Determine Whether a Relation Expressed as a Graph Represents a Function
4. Find the Value of a Function
5. Find the Domain of a Function
6. Work with Applications of Functions

Section 8.3 – Objective 1: Determine Whether a Relation Expressed as a Map or Ordered Pairs Represents a Function
Video Length – 10:01

In the last section, we learned that a relation is nothing more than a correspondence between two sets.

We called the set of _____ the _____ of the relation and we called the

corresponding set of all _____ the _____ of the relation.

Now we will look at a special type of relation called a **function**.

Definition
A _____ is a relation in which each element in the domain (the inputs) of the relation

corresponds to _____ _____ element in the range (the outputs) of the relation.

For example,

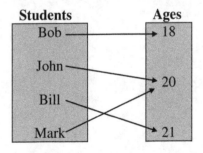

This is _____

_____

Course: Name:
Instructor: Section:

1. **Example:** Determine if the following relation represents a function.

   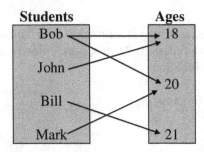

   Final answer: _____

   _____

2. **Example:** Determine if each relation represents a function. If it is a function, state the domain and range.

   (a) $\{(1,3),\ (0,5),\ (-5,2),\ (1,8),\ (10,-5)\}$

   Final answer: _____
   Note: Listen very VERY carefully to the justification he provides for the answer.

   (b) $\{(1,2),\ (0,5),\ (-5,2),\ (11,6),\ (9,-4),\ (-2,0)\}$

   Final answer: _____

Course: Name:
Instructor: Section:

Section 8.3 – Objective 2: Determine Whether a Relation Expressed as an Equation Represents a Function
Video Length – 5:41

Now we know how to determine whether a relation expressed as a map is a function. We know how to determine whether a relation expressed as ordered pairs is a function. Next we are going to look at determining whether a relation expressed as an equation is a function.

3. **Example:** Determine if each equation represents y as a function of x.

  (a) $y = -\dfrac{2}{3}x + 2$

  Final answer: _____

  _____

  (b) $y = \pm x^2 - 6$

  Final answer: _____

  _____

  Note: Make sure you provide justification for your answer. As always, use complete and coherent sentences.

Course:
Instructor:
Name:
Section:

Section 8.3 – Objective 3: Determine Whether a Relation Expressed as a Graph Represents a Function
Video Length – 7:14

We have identified functions from relations that were represented via a map, a set of ordered pairs, and by an equation. We also know that relations can be represented graphically. In order to understand how to identify a function from a relation that is presented as a graph, consider the following:

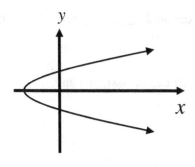

Vertical Line Test
A set of points in the *xy*-plane is the graph of a function if and only if _____ _____ intersects the graph in _____ _____ _____ point.

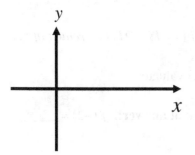
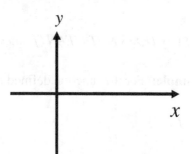

Note: The vertical line test can be broken up into two statements.

If a set of points in the *xy*-plane is the graph of a function, then _____
_____

If every vertical line intersects the graph in at most one point, then _____
_____

Course: Name:
Instructor: Section:

Section 8.3 – Objective 4: Find the Value of a Function
Part I – Text Example 6
Video Length – 8:47

Now that we know how to identify whether or not a relation is a function, we will now introduce you to function notation.

Functions are often denoted by letters such as f, g, F, G, and so on. If f is a function, then for each number x in its domain, the corresponding value in the range is denoted _____, read "f of x" or "f at x."

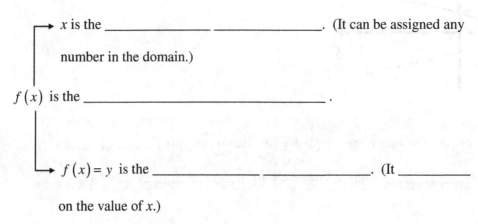

x is the _____ _____. (It can be assigned any number in the domain.)

$f(x)$ is the _____.

$f(x) = y$ is the _____ _____. (It _____ on the value of x.)

Note: $f(x)$ DOES NOT MEAN "f times x." In other words, $f(x)$ IS NOT the product of f and x.

4. **Example:** For the function defined by $f(x) = x^2 - 3x - 1$, evaluate:

(a) $f(-2)$ (a) **Final answer:** $f(-2) =$ _____

(b) $f(5)$ (b) **Final answer:** $f(5) =$ _____

Note: Go ahead and complete the next problem. Hint: You will get another quadratic function of x with positive coefficients.

(c) $f(x+4)$ (c) **Final answer:** $f(x+4) =$ _____

Course: Name:
Instructor: Section:

Section 8.3 – Objective 4: Find the Value of a Function
Part II – Text Example 7
Video Length – 9:16

Consider the function $G(a) = 3 - 2a$.

What is the name of the function? _____

What is the independent variable? _____

$G(a)$ is the _____ if the input is a.

The rule to get from the input to the output is _____.

5. **Example:** If $G(a) = 3 - 2a$, evaluate:

(a) $G(-10)$ (a) **Final answer:** $G(-10) =$ _____

(b) $G(a+2)$ (b) **Final answer:** $G(a+2) =$ _____

(c) $G(a) + G(2)$ (c) **Final answer:** $G(a) + G(2) =$ _____

Course:
Instructor:
Name:
Section:

Section 8.3 – Objective 5: Find the Domain of a Function
Video Length – 5:56

> **Definition**
> When only the equation of a function is given, we agree that the _____ of *f* is the largest set of real numbers for which $f(x)$ is a real number.

6. **Example:** Find the domain of the functions:

   (a) $h(x) = x^2 - 4$

   Final answer: _____

   (b) $f(x) = \dfrac{12}{x-5}$

   Final answer: _____

Course: Name:
Instructor: Section:

Section 8.3 – Objective 6: Work with Applications of Functions
Video Length – 4:21

There are many instances where functions play a role in our lives. For example, consider the following function.

$$N(h) = 15h - 0.2(15h - 100)$$

Note: Listen carefully to the description of each of the different parts of the function above. Write down these descriptions in the blanks below. (The answers to the following may not be presented in the suggested order.)

What does $N(h)$ represent? _____

_____

What does $15h$ represent? _____

_____

What does $(15h - 100)$ represent? _____

_____

What does $0.2(15h - 100)$ represent? _____

_____

What does N represent? _____

_____

What is the independent variable? What does it represent? _____

_____

Find $N(20)$ and describe, in words, what this represents.

- Find $N(20)$.

- What does $N(20)$ represent? _____

_____

Course:
Instructor:

Name:
Section:

Section 8.4 Video Guide
Functions and Their Graphs

Objectives:
1. Graph a Function
2. Obtain Information from the Graph of a Function
3. Know Properties and Graphs of Basic Functions
4. Interpret Graphs of Functions

Section 8.4 – Objective 1: Graph a Function
Video Length – 7:55

Graph the equation $y = 2x - 5$.

Does the graph of this equation pass the Vertical Line Test? _____ . Therefore, $y = 2x - 5$ represents a _____ .

Definition
A _____ **of a function** is the graph of the corresponding equation that defines the function.

When a function is defined by an equation, the graph of the function is _____
_____ .

1. **Example:** Graph $f(x) = |x| + 1$.

Course: Name:
Instructor: Section:

Section 8.4 – Objective 2: Obtain Information from the Graph of a Function
Part I – The Zero of a Function
Video Length – 10:21

2. **Example:** Determine the domain and range of the function. Identify the intercepts.

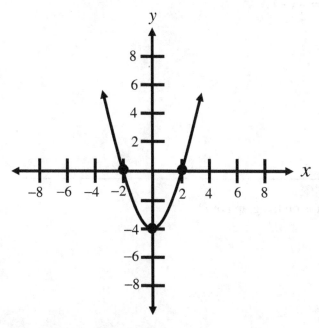

Final answer: Domain: _____ Range: _____

Intercepts: _____

3. **Example:** A young boy is swinging on a swing. In the graph below, let h be the distance he is above ground as a function of time t (in seconds).

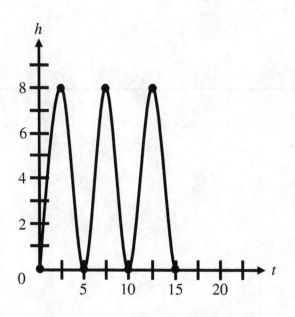

(a) What are $h(2.5)$ and $h(5)$? Interpret these values.

Final answer: _____

(b) What is the domain of h? What is the range of h?

Final answer: _____

(c) For what values of t does $h(t) = 8$?

Final answer: _____

Copyright © 2014 Pearson Education, Inc.

Course: Name:
Instructor: Section:

Section 8.4 – Objective 2: Obtain Information from the Graph of a Function
Part II – Text Examples 2 and 3
Video Length – 5:53

4. **Example:** Consider the function $f(x) = \dfrac{3}{2}x - 1$.

   (a) Is the point $(-2, -4)$ on the graph of f?

   Final answer: _____

   (b) If $x = 4$, what is $f(x)$? What point is on the graph of f?

   Final answer: _____

   (c) If $f(x) = 8$, what is x? What point is on the graph of f?
   Note: Make sure you know the difference in what is being asked between parts (b) and (c).

   Final answer: _____

Course: Name:
Instructor: Section:

Section 8.4 – Objective 2: Obtain Information from the Graph of a Function
Part III – Text Example 4
Video Length – 6:24

Definition
The _____ of a function is any number that causes the value of the function to equal 0.

In other words, _____.

Note: In the following example, make sure you note the relationship between the zeros of a function and a particular intercept of the graph of the function.

5. **Example:** Consider the function $f(x) = x^2 - 9$. Is 3 a zero of f?

   Final answer: _____

6. **Example:** What are the zeros of f?

   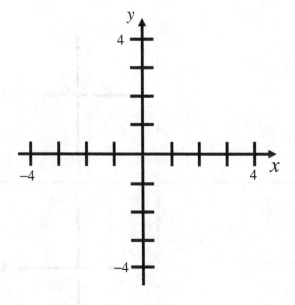

   Final answer: _____

Course: Name:
Instructor: Section:

Section 8.4 – Objective 3: Know Properties and Graphs of Basic Functions
Video Length – 3:32

Now we are going to talk about the Library of Functions. The Library of Functions represents a set of simple functions whose properties you should readily know.

Function	Properties	Graph
Linear Function		
Square Function		
Cube Function		
Absolute Value Function		

Course: Name:
Instructor: Section:

Section 8.4 – Objective 4: Interpret Graphs of Functions
Video Length – 3:48

We can use the graph of a function to give a visual description of many different scenarios. Consider the following example.

7. **Example:** Kevin decides to take a walk. He leaves his apartment and walks 1 block in 1 minute at a constant speed. He then jogs 4 blocks in 1 minute, walks 2 blocks in 3 minutes, and then sprints 3 blocks in 1 minute. At that point, Kevin rests for 1 minute before running home in 3 minutes. Draw a graph of Kevin's distance from home (in blocks) as a function of time (in minutes).

Course: Name:
Instructor: Section:

Section 8.5 Video Guide
Linear Functions and Models

Objectives:
1. Graph Linear Functions
2. Find the Zero of a Linear Function
3. Build Linear Models from Verbal Descriptions
4. Build Linear Models from Data

Section 8.5 – Objective 1: Graph Linear Functions
Video Length – 8:38

A linear equation is any equation of the form _____ . Consider the graphs of the four lines below.

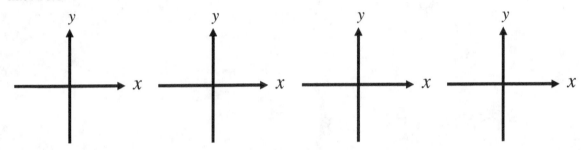

All linear equations, with the exception of those of the form _____ (_____ lines), are going to be functions. Which means we should be able to write them in function notation. In the linear equation $Ax + By = C$, we will treat x as the independent variable and y as the dependent variable. So we can rewrite a linear equation in function form as follows:

Definition
A _____ _____ is a function of the form

where m and b are real numbers. The graph of a linear function is called a _____ .

1. **Example:** Graph: $H(x) = -\dfrac{2}{3}x + 5$

Course:
Instructor:

Name:
Section:

Section 8.5 – Objective 2: Find the Zero of a Linear Function
Part I – Text Example 2
Video Length – 2:39

If r is a zero of the function f, then _____ .

2. **Example:** Find the zero of the function $g(x) = -\frac{2}{3}x + 4$.

Final answer: _____

Course:
Instructor:
Name:
Section:

Section 8.5 – Objective 2: Find the Zero of a Linear Function
Part II – Text Example 3
Video Length – 7:34

3. **Example:** The linear function $C(x) = 35 + 0.25x$ describes the cost for renting a car for x miles.

   (a) What is the implied domain of this function?

   Final answer: _____

   (b) What is $C(0)$? What does this result mean?

   Final answer: _____

   (c) What is the cost for renting a car and driving 500 miles?

   Final answer: _____

   (d) How many miles can you drive the rental car if you have $75?

   Final answer: _____

Course: Name:
Instructor: Section:

Section 8.5 – Objective 3: Build Linear Models from Verbal Descriptions
Part I – Text Example 4
Video Length – 6:34

> **Modeling with a Linear Function**
> If the average rate of change of a function is a constant m, a linear function f can be used to model the relation between the two variables as follows:
>
>
> where b is the value of f at 0. That is, _____ .

4. **Example:** Suppose a cab charges a flat rate of $3 plus $0.25 per mile.

   (a) Write a function that describes the cost C of traveling m miles.

   Final answer: _____

   (b) What is the cost of traveling 7.5 miles?

   Final answer: _____

   (c) If your bill was $8, how far did you travel?

   Final answer: _____

Course: Name:
Instructor: Section:

Section 8.5 – Objective 3: Build Linear Models from Verbal Descriptions
Part II – Text Example 5
Video Length – 9:12

5. **Example:** Suppose that a company just purchased some new office equipment at a cost of $2500 per machine. The company chooses to depreciate each machine using the straight-line method over 5 years.

(a) Build a linear model that expresses the value of V of each machine as a function of its age, x.

Final answer: _____

(b) What is the implied domain of the function found in part (a)?

Final answer: _____

(c) Graph the linear function.

(d) What is the value of the machine after 2 years?

Final answer: _____

(e) Interpret the slope.

Final answer: _____

(f) When will the value of each machine be $500?

Final answer: _____

Course: Name:
Instructor: Section:

Section 8.5 – Objective 4: Build Linear Models from Data
Video Length – 16:09

Often we are interested in finding an equation that describes the relation between two variables. To find this relation, a _____ _____ is used to plot the ordered pairs that make up the relation in the Cartesian plane.

6. **Example:** The data listed below represent the number of absences, x, and exam score, y.

   (a) Draw a scatter diagram treating the number of absences as the independent variable.

Number of Absences, x	Exam Score, y
0	89.2
1	86.4
2	83.5
2	81.1
4	78.2
4	73.9

   (b) Find a linear function that describes the relation between the number of absences, x, and the exam score, y.

   Final answer: _____

Course: Name:
Instructor: Section:

Section 8.6 Video Guide
Compound Inequalities

Objectives:
1. Determine the Intersection or Union of Two Sets
2. Solve Compound Inequalities Involving "and"
3. Solve Compound Inequalities Involving "or"
4. Solve Problems Using Compound Inequalities

Section 8.6 – Objective 1: Determine the Intersection or Union of Two Sets
Part I
Video Length – 3:18

We will now study compound inequalities.

Definition
The _____ of two sets A and B, denoted _____, is the set of all elements that belong to both set A _____ set B.

Let $A = \{1,2,3,4,5\}$ and let $B = \{4,5,6,7,8\}$.

So $A \cap B = $ _____

Definition
The _____ of two sets A and B, denoted _____, is the set of all elements that are in the set A _____ in the set B or in both A and B.

Let $A = \{1,2,3,4,5,6\}$ and let $B = \{2,4,6,8,10\}$.

So $A \cup B = $ _____

Course:
Instructor:
Name:
Section:

Section 8.6 – Objective 1: Determine the Intersection or Union of Two Sets
Part II – Example 2
Video Length – 11:57

1. **Example:** Let $A = \{x \mid x \geq 4\}$ and $B = \{x \mid x < 1\}$.

    (a) Graph sets A and B on a number line.

    Graph: ⎯⎯⎯⎯⎯⎯⎯⎯⎯▶

    Graph: ⎯⎯⎯⎯⎯⎯⎯⎯⎯▶

    (b) Find $A \cap B$ and $A \cup B$. Write solutions in interval notation.

    Graph: ⎯⎯⎯⎯⎯⎯⎯⎯⎯▶   **Interval notation:** ⎯⎯⎯⎯⎯⎯⎯⎯

    Graph: ⎯⎯⎯⎯⎯⎯⎯⎯⎯▶   **Interval notation:** ⎯⎯⎯⎯⎯⎯⎯⎯

2. **Example:** Let $A = \{x \mid x \leq 5\}$ and $B = \{x \mid x > -1\}$.

    (a) Find $A \cup B$.

    Graph: ⎯⎯⎯⎯⎯⎯⎯⎯⎯▶

    Set-builder: ⎯⎯⎯⎯⎯⎯⎯⎯⎯

    Interval notation: ⎯⎯⎯⎯⎯⎯⎯⎯⎯

    (b) Find $A \cap B$.

    Graph: ⎯⎯⎯⎯⎯⎯⎯⎯⎯▶

    Set-builder: ⎯⎯⎯⎯⎯⎯⎯⎯⎯

    Interval notation: ⎯⎯⎯⎯⎯⎯⎯⎯⎯

Course: Name:
Instructor: Section:

Section 8.6 – Objective 2: Solve Compound Inequalities Involving "and"
Part I
Video Length – 7:27

Definition

A _____ _____ is formed by joining two inequalities with the word "_____"

or "_____".

3. **Example:** Solve the compound inequality and graph the solution set.

       $x + 3 \leq 7$ *and* $x - 2 \leq -3$

       Graph: _____⟶

       Set-builder: _____

       Interval notation: _____

Note: Remember, be careful when you divide (or multiply) an inequality by a negative number.

4. **Example:** Solve: $3x + 1 > 4$ and $-2x + 5 \geq -7$

       Graph: _____⟶

       Set-builder: _____

       Interval notation: _____

Steps for Solving Compound Inequalities Involving "and"

Step 1: Solve each inequality separately.

Step 2: Find the _____ of the solution sets of each inequality.

Course: Name:
Instructor: Section:

Section 8.6 – Objective 2: Solve Compound Inequalities Involving "and"
Part II – Text Example 6
Video Length – 3:58

Sometimes "and" inequalities can be written in shorthand. For example,

$a < x$ and $x < b$

5. **Example:** Solve the compound inequality.

   $-3 < -4x + 1 < 17$

   Set-builder: _____

   Interval notation: _____

   Graph: ───────────────▶

Course:
Instructor:

Name:
Section:

Section 8.6 – Objective 3: Solve Compound Inequalities Involving "or"
Video Length – 5:34

6. **Example:** Solve the compound inequality and graph the solution set.

   $6x - 10 < 8$ *or* $2x + 1 > 9$

   Set-builder: _____

   Interval notation: _____

   Graph: ────────────▶

7. **Example:** Solve the compound inequality.

   $3x + 1 \leq 7$ *or* $-4x + 1 \geq 5$

   Graph: ────────────▶

   Set-builder: _____

   Interval notation: _____

Course: Name:
Instructor: Section:

Section 8.6 – Objective 4: Solve Problems Using Compound Inequalities
Video Length – 9:50

8. **Example:** Joanna desperately wants to earn a *B* in her History class. Her current test scores are 74, 86, 77, and 89. Her final is worth 2 test scores. In order to earn a *B*, Joanna's average must like between 80 and 89, inclusive. What range of scores can Joanna receive on the final and earn a *B* in the course?

Final answer: _____

Course: Name:
Instructor: Section:

Section 8.7 Video Guide
Absolute Value Equations and Inequalities

Objectives:
1. Solve Absolute Value Equations
2. Solve Absolute Value Inequalities Involving < or ≤
3. Solve Absolute Value Inequalities Involving > or ≥
4. Solve Applied Problems Involving Absolute Value Inequalities

Section 8.7 – Objective 1: Solve Absolute Value Equations
Part I – Text Examples 1 and 2
Video Length – 11:38

Recall, $|x|$ is the _____

So, $|x| = 5$ is saying, "_____

_____."

So the solution set of the equation $|x| = 5$ is _____ .

Graph $f(x) = |x|$.

Equations Involving Absolute Value
If a is a _____ real number and if u is any algebraic expression, then

$$|u| = a \text{ is equivalent to } _____ \text{ or } _____ .$$

Note: If _____ , the equation _____ is equivalent to _____ . If _____ , the equation _____ has ____ _____ _____ .

Course: Name:
Instructor: Section:

Steps for Solving Absolute Value Equations with One Absolute Value

Step 1: Isolate the expression containing the absolute value.

Step 2: Rewrite the absolute value equation as two equations: _____ and _____ , where u is the algebraic expression in the absolute value symbol.

Step 3: Solve each equation.

Step 4: _____ your solution.

Here we go!

1. **Example:** Solve the equation $|c - 15| = 23$.

Write the steps in words	Show the steps with math
Step 1	
Step 2	
Step 3	
Step 4	

Final answer: _____

Course: Name:
Instructor: Section:

Section 8.7 – Objective 1: Solve Absolute Value Equations
Part II – Text Example 3
Video Length – 1:14

2. **Example:** Solve the equation $|5 - x| + 8 = 1$.

Final answer: _____

Course: Name:
Instructor: Section:

Section 8.7 – Objective 1: Solve Absolute Value Equations
Part III – Text Example 4
Video Length – 8:09

Equations Involving Absolute Values
If u and v are any algebraic expressions, then

$$|u| = |v| \text{ is equivalent to } \underline{\hspace{1cm}} \text{ or } \underline{\hspace{1cm}}.$$

3. **Example:** Solve the equation $|5b + 3| = |12 - 4b|$.

Final answer: _____

Course: Name:
Instructor: Section:

Section 8.7 – Objective 2: Solve Absolute Value Inequalities Involving < or ≤
Video Length – 9:37

Recall, $|x| < 3$ is saying "_____

_____."

⟶

So the graph above in inequality form is _____ .

Graph $f(x) = |x|$ and $g(x) = 3$.

Inequalities of the Form < or ≤ Involving Absolute Value
If a is a _____ real number and if u is an algebraic expression, then

$|u| < a$ is equivalent to _____

$|u| \leq a$ is equivalent to _____ .

Note: If _____ , _____ has _____ _____ _____ , _____ is equivalent to

_____ . If _____ , the inequality has _____ _____ _____ .

---

4. **Example:** Solve the inequality $|y - 15| < 23$. Write the solution set using set-builder notation and using interval notation.

   Set-builder: _____

   Interval notation: _____

306   Copyright © 2014 Pearson Education, Inc.

Course: Name:
Instructor: Section:

Section 8.7 – Objective 3: Solve Absolute Value Inequalities Involving > or ≥
Video Length – 12:24

Recall, $|x| > 4$ is saying "_____

_____."

→

So the graph above in inequality form is _____ .

Graph $f(x) = |x|$ and $g(x) = 4$.

Inequalities of the Form > or ≥ Involving Absolute Value
If a is a _____ real number and if u is an algebraic expression, then

$|u| > a$ is equivalent to _____ or _____

$|u| \geq a$ is equivalent to _____ or _____ .

5. **Example:** Solve: $|-w+3| - 5 \geq 2$

Set-builder: _____

Interval notation: _____

Graph: →

Course: Name:
Instructor: Section:

Section 8.7 – Objective 4: Solve Applied Problems Involving Absolute Value Inequalities
Video Length – 2:29

You may have read phrases such as "margin of error" and "tolerance." These are examples that lead to absolute value inequalities.

6. **Example:** The inequality $|x - 98.6| \geq 1.5$ represents a human body temperature (measured in degrees Fahrenheit) that is considered "unhealthy." Solve the inequality and interpret the results.

Final answer: _____

Course: Name:
Instructor: Section:

Section 8.8 Video Guide
Variation

Objectives:
1. Model and Solve Direct Variation Problems
2. Model and Solve Inverse Variation Problems
3. Model and Solve Joint and Combined Variation Problems

Section 8.8 – Objective 1: Model and Solve Direct Variation Problems
Video Length – 6:06

Definition
_____ describes how one quantity changes in relation to another quantity.

Definition
If x and y represent two quantities, we say that y _____ _____ with x, or

y is _____ _____ _____ x, if there is a nonzero number k such that

$$\frac{\quad}{\quad} = \underline{\quad\quad}.$$

_____ ____ _____

(_____ ____ _____)

1. **Example:** Suppose that y varies directly with x, for $x \geq 0$. Find an equation that relates y and x if it is known that $y = 195$ when $x = 15$.

   Final answer: _____

2. **Example:** The Smith family just purchased a house for $120,000, using a 15-year mortgage at 5.5% interest. Their monthly payments are $980.50. The monthly payment p varies directly with the amount borrowed b. If the Smiths decided to buy a more expensive home, what would the monthly payments be for a $150,000 mortgage?

   Final answer: _____

Copyright © 2014 Pearson Education, Inc.

309

Course: Name:
Instructor: Section:

Section 8.8 – Objective 2: Model and Solve Inverse Variation Problems
Video Length – 5:16

We are now going to look at **inverse variation**.

Definition
If x and y represent two quantities, we say that y _____ _____ with x, or y is _____ _____ _____ x, if there is a nonzero number k such that

___ = ___ .

3. **Example:** Suppose that y varies inversely with x, for $x \geq 0$. Find an equation that relates y and x if it is known that $y = 24$ when $x = 10$.

   Final answer: _____

4. **Example:** The frequency of sound varies inversely with the wave length. If a radio station broadcasts at a frequency of 90 megahertz, the wave length of sound is approximately 3 meters. Find the wave length of a radio broadcast at 20 megahertz.

   Final answer: _____

Course: Name:
Instructor: Section:

Section 8.8 – Objective 3: Model and Solve Joint and Combined Variation Problems
Video Length – 5:45

Definition
When a variable quantity Q is proportional to the product of two or more other variables, we say that

Q _____ _____ with these quantities.

For example, the equation $y = kxz$ can be read "_____."

Definition
When direct and inverse variation occur at the same time, we have _____

_____.

For example, the equation $y = \dfrac{kx}{z}$ can be read "_____."

5. **Example:** The kinetic energy K of a moving object varies jointly with its mass m and the square of its velocity v. If an object weighing 25 kilograms and moving at a velocity of 10 meters per second has a kinetic energy of 1250 joules, find its kinetic energy when the velocity is 15 meters per second.

   Final answer: _____

6. **Example:** A person's body mass index (BMI) is used by medical professionals. A BMI of 19 to 25 is considered desirable. BMI varies directly with the weight (in pounds) of a person and inversely with the square of the height (in inches). If a person who is 5 feet tall and weighs 105 pounds has a BMI of 20.5, what is the BMI of a person who is 5 feet, 8 inches tall and weighs 145 pounds?

   Final answer: _____

Course: Name:
Instructor: Section:

Course: Name:
Instructor: Section:

Section 9.1 Video Guide
Radicals and Rational Exponents

Objectives:
1. Evaluate Square Roots of Perfect Squares
2. Determine Whether a Square Root Is Rational, Irrational, or Not a Real Number
3. Find Square Roots of Variable Expressions

Section 9.1 – Objective 1: Evaluate Square Roots of Perfect Squares
Video Length – 8:41

Note: In this section, the "undoing" or "reversing" of the squaring process will be introduced.

Definition
For any real numbers a and b, b is a _____ _____ of a if ____ = ____ .

For example, because $3^2 = 9$, then 3 is a square root of 9 (*Note: –3 is also a square root of 9*).

What are the square roots of 16? _____ .

There are actually _____ results.

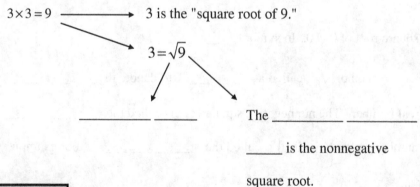

$3 \times 3 = 9$ ⟶ 3 is the "square root of 9."

$3 = \sqrt{9}$

_____ ____ The _____ _____

_____ is the nonnegative

square root.

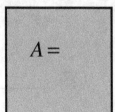

$A =$

Note: The side length of this square represents the principal square root of the area.

Course:
Instructor:

Name:
Section:

Find the square roots of 49.

What is the positive square root of 49?

What are the square roots of $\dfrac{9}{100}$?

What is the positive square root of $\dfrac{9}{100}$?

Properties of Square Roots

- Every positive real number has _____ _____ _____, one positive and one negative.

- The square root of 0 is 0. In symbols, ____ = ____.

- We use the symbol $\sqrt{}$, called a _____, to denote the _____ square root of a real number. The nonnegative square root is called the _____ square root.

- The number under the radical is called the _____. For example, the radicand in $\sqrt{25}$ is _____.

1. **Example:** Evaluate each square root:

   (a) $\sqrt{121}$ (a) **Final answer:** $\sqrt{121} =$ _____

   (b) $\sqrt{\dfrac{16}{49}}$ (b) **Final answer:** $\sqrt{\dfrac{16}{49}} =$ _____

   (c) $-\sqrt{100}$ (c) **Final answer:** $-\sqrt{100} =$ _____

Course: Name:
Instructor: Section:

Section 9.1 – Objective 2: Determine Whether a Square Root Is Rational, Irrational, or Not a Real Number

Video Length – 10:12

We are now going to look at some additional properties of square roots.

More Properties of Square Roots

- The square root of a perfect square is a _____ number.

- The square root of a positive rational number that is not a perfect square is an _____ number. For example, $\sqrt{20}$ is an irrational number because 20 is not a perfect square.

- The square root of a negative real number is _____ a _____ number. For example, $\sqrt{-2}$ is not a real number.

---

2. **Example:** Approximate $\sqrt{17}$ by writing it rounded to two decimal places.

   Final answer: _____

3. **Example:** Determine if each square root is rational, irrational, or not a real number:

   (a) $\sqrt{11}$ (a) **Final answer:** _____

   (b) $\sqrt{144}$ (b) **Final answer:** _____

   (c) $\sqrt{-54}$ (c) **Final answer:** _____

Note: Remember, the radicand is the number (or expression) under the radical.

Course: Name:
Instructor: Section:

Section 9.1 – Objective 3: Find Square Roots of Variable Expressions
Video Length – 6:23

Find the following.

$$\sqrt{3^2}$$

$$\sqrt{(-3)^2}$$

What about the following? *Note: Be careful with this one.*

$$\sqrt{x^2}$$

Definition
For any real number a,

_____ = _____ .

4. **Example:** Simplify the following.

(a) $\sqrt{(81x)^2}$ (a) **Final answer:** $\sqrt{(81x)^2} =$ _____

(b) $\sqrt{(a+4)^2}$ (b) **Final answer:** $\sqrt{(a+4)^2} =$ _____

Note: Observe the following restriction on the given variable and how it affects the final answer.

(c) $\sqrt{x^2}$, $x > 0$ (c) **Final answer:** $\sqrt{x^2} =$ _____

(d) $\sqrt{(p-2)^2}$, $p \geq 2$ (d) **Final answer:** $\sqrt{(p-2)^2} =$ _____

Course: Name:
Instructor: Section:

Section 9.2 Video Guide
nth Roots and Rational Exponents

Objectives:
1. Evaluate nth Roots
2. Simplify Expressions of the Form $\sqrt[n]{a^n}$
3. Evaluate Expressions of the Form $a^{\frac{1}{n}}$
4. Evaluate Expressions of the Form $a^{\frac{m}{n}}$

Section 9.2 – Objective 1: Evaluate nth Roots
Video Length – 7:35

Definition
The _____ ____ _____ **of a number a**, symbolized by _____ , where $n \geq 2$ is an integer, is defined as follows:

_____ = ____ means ____ = _____ .

In other words, the nth root of some number a means that _____

_____

_____

For example, if you're asked to find $\sqrt[3]{8}$, ask yourself _____

_____

Note: He doesn't actually write down the meaning of $\sqrt[4]{16}$, but he says it. Make sure YOU write it down.

If you're asked to find $\sqrt[4]{16}$, what do you ask yourself? What number _____

_____

Note: Pay attention to what he says about the 'index'.

1. **Example:** Evaluate:

    (a) $\sqrt{121}$ (a) **Final answer:** $\sqrt{121} =$ _____

    (b) $\sqrt[3]{-1000}$ (b) **Final answer:** $\sqrt[3]{-1000} =$ _____

Course: Name:
Instructor: Section:

The square root of a negative number _____

_____

The fourth root of a negative number _____

For example, we know the fourth root of a negative number won't be _____ because

_____

The fourth root of a negative number is not _____ either because _____

_____

Consider $\sqrt[n]{a}$. If n is _____, then _____. If n is _____, then _____.

For example, $\sqrt[4]{-12}$ is _____

Course:
Instructor:

Name:
Section:

Section 9.2 – Objective 2: Simplify Expressions of the Form $\sqrt[n]{a^n}$
Video Length – 9:34

Simplifying $\sqrt[n]{a^n}$
If $n \geq 2$ is a positive integer and a is a real number, then

_____ = _____ if $n \geq 3$ is _____

_____ = _____ if $n \geq 2$ is _____ .

For example:

2. **Example:** Simplify:

(a) $\sqrt[4]{x^4}$

(a) **Final answer:** $\sqrt[4]{x^4}$ = _____

(b) $\sqrt[3]{-27a^6}$

(b) **Final answer:** $\sqrt[3]{-27a^6}$ = _____

(c) $\sqrt[4]{16y^8}$

(c) **Final answer:** $\sqrt[4]{16y^8}$ = _____

Course:
Instructor:
Name:
Section:

Section 9.2 – Objective 3: Evaluate Expressions of the Form $a^{\frac{1}{n}}$
Video Length – 7:59

Up to this point, we have only worked with integer exponents. Now what we want to do is to define what it means for an expression to have a rational exponent.

Consider

$$\left(8^{1/2}\right)^2 \qquad\qquad \left(\sqrt{8}\right)^2$$

We can conjecture that _____ = _____ .

Definition of $a^{1/n}$
If a is a real number and n is an integer with $n \geq 2$, then

_____ = _____ .

Basically, the denominator of the rational exponent becomes _____.

3. **Example:** Write each of the following expressions as a radical and simplify, if possible.

   (a) $16^{1/2}$ (a) **Final answer:** $16^{1/2} =$ _____

   (b) $w^{1/4}$ (b) **Final answer:** $w^{1/4} =$ _____

   (c) $(3abc)^{1/5}$ (c) **Final answer:** $(3abc)^{1/5} =$ _____

4. **Example:** Rewrite each of the following radicals with a rational exponent.

   (a) $\sqrt[3]{6x}$ (a) **Final answer:** $\sqrt[3]{6x} =$ _____

   (b) $\sqrt[5]{(12c)^3}$ (b) **Final answer:** $\sqrt[5]{(12c)^3} =$ _____

   (c) $\sqrt[6]{\left(\dfrac{2x}{y}\right)^5}$ (c) **Final answer:** $\sqrt[6]{\left(\dfrac{2x}{y}\right)^5} =$ _____

Note: Part (d) is already written with a rational exponent. However, it can be simplified further.

   (d) $(-64)^{2/3}$ (d) **Final answer:** $(-64)^{2/3} =$ _____

Course: Name:
Instructor: Section:

Section 9.2 – Objective 4: Evaluate Expressions of the Form $a^{\frac{m}{n}}$
Video Length – 8:40

Definition of $a^{m/n}$
If a is a real number, m/n is a rational number in _____ _____ with $n \geq 2$, then

_____ = _____ = _____ = _____ = _____

provided that $\sqrt[n]{a}$ exists.

In other words, when you have $a^{m/n}$, the denominator of the rational exponent _____

_____

5. **Example:** Evaluate each of the following expressions, if possible.

    (a) $16^{2/3}$ (a) **Final answer:** $16^{2/3} =$ _____

    (b) $w^{3/4}$ (b) **Final answer:** $w^{3/4} =$ _____

    (c) $(3abc)^{2/5}$ (c) **Final answer:** $(3abc)^{2/5} =$ _____

Negative Exponent Rule
If $\frac{m}{n}$ is a rational number, and if a is a nonzero number, then we define

_____ = _____ and _____ = _____ if $a \neq 0$.

6. **Example:** Rewrite each of the following with positive exponents and completely simplify, if possible.

    (a) $9^{-\frac{1}{2}}$ (a) **Final answer:** $9^{-\frac{1}{2}} =$ _____

    (b) $\dfrac{1}{25^{-\frac{3}{2}}}$ (b) **Final answer:** $\dfrac{1}{25^{-\frac{3}{2}}} =$ _____

Copyright © 2014 Pearson Education, Inc.

Course: Name:
Instructor: Section:

Section 9.3 Video Guide
Simplify Expressions Using the Law of Exponents

Objectives:
1. Simplify Expressions Involving Rational Exponents
2. Simplify Radical Expressions
3. Factor Expressions Containing Rational Exponents

Section 9.3 – Objective 1: Simplify Expressions Involving Rational Exponents
Video Length – 15:23

The Law of Exponents	
If a and b are real numbers and if r and s are rational numbers, then assuming the expression is defined,	
Zero-Exponent Rule:	
Negative-Exponent Rule:	
Product Rule:	
Quotient Rule:	
Power Rule:	
Product to Power Rule:	
Quotient to Power Rule:	
Quotient to a Negative Power Rule:	

Definition
The direction **simplify** shall mean the following:

- _____ _____ are _____.

- Each _____ occurs _____.

- There are no _____ in the expression.

- There are no _____ written to _____.

Course:
Instructor:

Name:
Section:

1. **Example:** Simplify the following:

    (a) $9^{1/4} \cdot 9^{1/3}$

    Final answer: $9^{1/4} \cdot 9^{1/3} =$ _____

    (b) $\dfrac{y^{1/5}}{y^{9/10}}$

    Final answer: $\dfrac{y^{1/5}}{y^{9/10}} =$ _____

2. **Example:** Simplify the following:

    (a) $\left(y^{-4/5}\right)^{1/3}$

    Final answer: $\left(y^{-4/5}\right)^{1/3} =$ _____

    Note: He skips part (b).
    (b) $-6a^{(-4/9)+(1/9)} + 3a^{(-4/9)+2}$

    Final answer: $-6a^{(-4/9)+(1/9)} + 3a^{(-4/9)+2} =$ _____

    (c) $\left(25x^{2/5}y^{-1}\right)^{1/2}$

    Final answer: $\left(25x^{2/5}y^{-1}\right)^{1/2} =$ _____

Copyright © 2014 Pearson Education, Inc.

Course:
Instructor:

Name:
Section:

3. **Example:** Simplify the following:

$$\left(\frac{64a^{1/2}b}{a^{-2}b^{3/4}}\right)^{1/2}$$

Final answer: $\left(\dfrac{64a^{1/2}b}{a^{-2}b^{3/4}}\right)^{1/2} = $ _____

Course: Name:
Instructor: Section:

Section 9.3 – Objective 2: Simplify Radical Expressions
Video Length – 2:36

4. **Example:** Simplify the following:

   (a) $\sqrt[3]{a^6}$

   Final answer: $\sqrt[3]{a^6} =$ _____

   (b) $\dfrac{\sqrt[3]{w^2}}{\sqrt{w}}$

   Final answer: $\dfrac{\sqrt[3]{w^2}}{\sqrt{w}} =$ _____

   Note: The video was cut a little short. However, the final answer does show up.

Course: Name:
Instructor: Section:

Section 9.3 – Objective 3: Factor Expressions Containing Rational Exponents
Video Length – 9:14

5. **Example:** Simplify $x^{1/4} - x^{5/4}$ by factoring out $x^{1/4}$.

   Final answer: $x^{1/4} - x^{5/4} = $ _____

6. **Example:** Simplify $3x^{\frac{3}{2}} + 2x^{\frac{1}{2}}(4x+1)$ by factoring out $x^{\frac{1}{2}}$.

   Final answer: $3x^{\frac{3}{2}} + 2x^{\frac{1}{2}}(4x+1) = $ _____

7. **Example:** Simplify $2x^{\frac{2}{3}} + x^{-\frac{1}{3}}(3x+2)$ by factoring out $x^{-\frac{1}{3}}$.

   Final answer: $2x^{\frac{2}{3}} + x^{-\frac{1}{3}}(3x+2) = $ _____

Course: Name:
Instructor: Section:

Section 9.4 Video Guide
Simplify Radical Expressions Using Properties of Radicals

Objectives:
1. Use the Product Property to Multiply Radical Expressions
2. Use the Product Property to Simplify Radical Expressions
3. Use the Quotient Property to Simplify Radical Expressions
4. Multiply Radicals with Unlike Indices

Section 9.4 – Objective 1: Use the Product Property to Multiply Radical Expressions
Video Length – 4:25

Product Property of Radicals
If $\sqrt[n]{a}$ and $\sqrt[n]{b}$ are real numbers and $n \geq 2$ is an integer, then

$$\underline{\hspace{2cm}} = \underline{\hspace{2cm}}.$$

Note: Remember, in order to multiply radicals, the index needs to be the same.

1. **Example:** Multiply:

   (a) $\sqrt[3]{2} \cdot \sqrt[3]{4}$

            Final answer: $\sqrt[3]{2} \cdot \sqrt[3]{4} =$ _____

   (b) $\sqrt{c+4} \cdot \sqrt{c-4}$

            Final answer: $\sqrt{c+4} \cdot \sqrt{c-4} =$ _____
                                                  Note: WARNING!

   (c) $\sqrt[5]{3x^2} \cdot \sqrt[5]{6x}$

            Final answer: $\sqrt[5]{3x^2} \cdot \sqrt[5]{6x} =$ _____

Course: Name:
Instructor: Section:

Section 9.4 – Objective 2: Use the Product Property to Simplify Radical Expressions
Part I – Text Examples 2, 3, 4, 5, and 6
Video Length – 19:08

Definition
A radical expression is _____ provided that the radicand does not contain any factors that are _____ of the _____ .

Simplifying a Radical Expression

Step 1: Write each factor of the radicand as the product of two factors, one of which is a

_____ _____ of the index.

Step 2: Write the radicand as the product of two radicals, one of which contains perfect squares.

Step 3: Take the *n*th root of each perfect power.

2. **Example:** Simplify the following:

   (a) $\sqrt{50}$

Write the steps in words	Show the steps with math
Step 1	
Step 2	
Step 3	

   Final answer: $\sqrt{50} =$ _____

   (b) $\sqrt{48x^{15}}$

   Final answer: $\sqrt{48x^{15}} =$ _____

Course: Name:
Instructor: Section:

3. **Example:** Simplify the following:

(a) $\sqrt{81c^8}$

Final answer: $\sqrt{81c^8} =$ _____

(b) $\sqrt[3]{x^2 y^4 z^{10}}$

Final answer: $\sqrt[3]{x^2 y^4 z^{10}} =$ _____

(c) $\sqrt[4]{32x^4}$

Final answer: $\sqrt[4]{32x^4} =$ _____

(d) $\dfrac{9+\sqrt{18}}{3}$

Final answer: $\dfrac{9+\sqrt{18}}{3} =$ _____

Course: Name:
Instructor: Section:

Section 9.4 – Objective 2: Use the Product Property to Simplify Radical Expressions
Part II – Text Example 7
Video Length – 10:29

4. **Example:** Simplify the following:

   (a) $2\sqrt[3]{9x^2} \cdot \sqrt[3]{3x^2}$

   Final answer: $2\sqrt[3]{9x^2} \cdot \sqrt[3]{3x^2} = $ _____

   Note: In the example above, he computes the product of 9 and 3 to get 27. He then writes $27 = 3^3$. But observe how he writes the product of 8 and 12 in the next example.

   (b) $\sqrt[4]{8m^3n^5} \cdot \sqrt[4]{12m^2n^5}$

   Final answer: $\sqrt[4]{8m^3n^5} \cdot \sqrt[4]{12m^2n^5} = $ _____

Course: Name:
Instructor: Section:

Section 9.4 – Objective 3: Use the Quotient Property to Simplify Radical Expressions
Part I – Text Example 8
Video Length – 6:49

Quotient Property of Radicals

If $\sqrt[n]{a}$ and $\sqrt[n]{b}$ are real numbers, $b \neq 0$, and $n \geq 2$ is an integer, then

$$\underline{} = \underline{}$$

5. **Example:** Simplify: $\sqrt{\dfrac{75}{81}}$

   Final answer: $\sqrt{\dfrac{75}{81}} = \underline{}$

6. **Example:** Simplify: $\sqrt{\dfrac{64x^5}{2x^3}}$

   Final answer: $\sqrt{\dfrac{64x^5}{2x^3}} = \underline{}$

7. **Example:** Simplify: $\sqrt{\dfrac{5pq^4}{9r^4}}$

   Final answer: $\sqrt{\dfrac{5pq^4}{9r^4}} = \underline{}$

Course: Name:
Instructor: Section:

Section 9.4 – Objective 3: Use the Quotient Property to Simplify Radical Expressions
Part II – Text Example 9
Video Length – 4:00

8. **Example:** Simplify: $\dfrac{\sqrt{45y^5}}{\sqrt{5y}}$

Final answer: $\dfrac{\sqrt{45y^5}}{\sqrt{5y}} =$ _____

9. **Example:** Simplify: $\dfrac{-2\sqrt[3]{250n}}{\sqrt[3]{2n^4}}$

Final answer: $\dfrac{-2\sqrt[3]{250n}}{\sqrt[3]{2n^4}} =$ _____

Course: Name:
Instructor: Section:

Section 9.4 – Objective 4: Multiply Radicals with Unlike Indices
Video Length – 3:48

10. Example: Multiply and simplify: $\sqrt[3]{4} \cdot \sqrt[4]{2}$

Final answer: $\sqrt[3]{4} \cdot \sqrt[4]{2} = $ _____

Course: Name:
Instructor: Section:

Section 9.5 Video Guide
Adding, Subtracting, and Multiplying Radical Expressions

Objectives:
1. Add or Subtract Radical Expressions
2. Multiply Radical Expressions

Section 9.5 – Objective 1: Add or Subtract Radical Expressions
Video Length – 8:35

Definition
Two radicals are _____ _____ if each radical has the same _____ and the same _____. We can add or subtract like radicals.

Essentially, adding or subtracting like radicals is just like combining like terms.

1. **Example:** Combine the following:

   (a) $5\sqrt[5]{2} + 4\sqrt[5]{2}$

   Final answer: $5\sqrt[5]{2} + 4\sqrt[5]{2} = $ _____

   (b) $3\sqrt{xyz^2} + 10\sqrt{xyz^2} - 5\sqrt{xyz^2}$

   Final answer: $3\sqrt{xyz^2} + 10\sqrt{xyz^2} - 5\sqrt{xyz^2} = $ _____

   (c) $6\sqrt[5]{7} + 7\sqrt[5]{6}$

   Final answer: $6\sqrt[5]{7} + 7\sqrt[5]{6} = $ _____

Course: Name:
Instructor: Section:

2. **Example:** Combine the following:

   (a) $\sqrt[3]{256} - 4\sqrt[3]{32}$

   Final answer: $\sqrt[3]{256} - 4\sqrt[3]{32} =$ _____

   (b) $3x\sqrt[3]{x^4 y} - 2\sqrt[3]{x^3 y^5}$

   Final answer: $3x\sqrt[3]{x^4 y} - 2\sqrt[3]{x^3 y^5} =$ _____

Course:
Instructor:
Name:
Section:

Section 9.5 – Objective 2: Multiply Radical Expressions
Part I – Text Example 4
Video Length – 6:02

We are now going to use the Distributive Property with radicals. Recall, the Distributive Property states

$$a(b+c)$$

3. **Example:** Multiply: $-\sqrt{3}\left(4-2\sqrt{3}\right)$

Final answer: $-\sqrt{3}\left(4-2\sqrt{3}\right) = $ _____

Recall the FOIL method for multiplying binomials. For example,

$$(x+3)(2x-5)$$

4. **Example:** Multiply:

(a) $\left(8+\sqrt{5}\right)\left(6-\sqrt{2}\right)$

Final answer: $\left(8+\sqrt{5}\right)\left(6-\sqrt{2}\right) = $ _____

(b) $\left(7-\sqrt{5w}\right)\left(2-\sqrt{5w}\right)$

Final answer: $\left(7-\sqrt{5w}\right)\left(2-\sqrt{5w}\right) = $ _____

Copyright © 2014 Pearson Education, Inc.

Course: Name:
Instructor: Section:

Section 9.5 – Objective 2: Multiply Radical Expressions
Part II – Text Example 5
Video Length – 8:11

Recall the following special product formulas. For example,

$(A+B)^2$

$(A-B)^2$

5. **Example:** Multiply:

  (a) $(9+\sqrt{3})^2$

  Final answer: $(9+\sqrt{3})^2 =$ _____

  (b) $(6-\sqrt{5})^2$

  Final answer: $(6-\sqrt{5})^2 =$ _____

Course:
Instructor:

Name:
Section:

Do you remember the formula for the difference of two squares? Recall,

$$(A+B)(A-B)$$

6. **Example:** Multiply: $\left(2-3\sqrt{5}\right)\left(2+3\sqrt{5}\right)$

Final answer: $\left(2-3\sqrt{5}\right)\left(2+3\sqrt{5}\right) = $ _____

Note: He makes a really good point about these special product formulas and FOIL. If you forget the formulas, it is not the end of the world. You can use FOIL.

Course: Name:
Instructor: Section:

Section 9.6 Video Guide
Rationalizing Radical Expressions

Objectives:
1. Rationalize a Denominator Containing One Term
2. Rationalize a Denominator Containing Two Terms

Section 9.6 – Objective 1: Rationalize a Denominator Containing One Term
Video Length – 10:09

Definition
When radical expressions appear in the denominator of a quotient, we need to rewrite the quotient so that the denominator does not contain radicals. This process is called _____ _____ .

Why is this called process called rationalizing the denominator? If there is a radical in the denominator that represents an irrational number, we would like to rewrite the expression so that the denominator is a rational number. For example,

$$\frac{1}{\sqrt{3}}$$

1. **Example:** Simplify: $\sqrt{\dfrac{5pq^4}{2r}}$

   Final answer: $\sqrt{\dfrac{5pq^4}{2r}} = $ _____

2. **Example:** Simplify: $\dfrac{6}{\sqrt[3]{9b^2}}$

   Final answer: $\dfrac{6}{\sqrt[3]{9b^2}} = $ _____

Course: Name:
Instructor: Section:

Section 9.6 – Objective 2: Rationalize a Denominator Containing Two Terms
Video Length – 9:47

We just learned how to rationalize a denominator that contains a single radical. Now we are going to learn how to rationalize a denominator contains two terms.

To rationalize a denominator containing two terms, we use the fact that

$$\underline{\hspace{2cm}} = \underline{\hspace{1cm}}$$

and multiply both numerator and denominator of the quotient by the _____ of the denominator.

For example, the conjugate of $5+\sqrt{6}$ is _____. The conjugate of $\sqrt{3x}-\sqrt{2y}$ is _____.

3. **Example:** Rationalize the denominator: $\dfrac{5}{\sqrt{2}+1}$

   Final answer: $\dfrac{5}{\sqrt{2}+1} = $ _____

340   Copyright © 2014 Pearson Education, Inc.

Course: Name:
Instructor: Section:

4. **Example:** Rationalize the denominator: $\dfrac{3\sqrt{6}+5\sqrt{7}}{2\sqrt{6}-3\sqrt{7}}$

Final answer: $\dfrac{3\sqrt{6}+5\sqrt{7}}{2\sqrt{6}-3\sqrt{7}} = $ _____

Course: Name:
Instructor: Section:

Section 9.7 Video Guide
Functions Involving Radicals

Objectives:
1. Evaluate Functions Involving Radicals
2. Find the Domain of a Function Involving a Radical
3. Graph Functions Involving Square Roots
4. Graph Functions Involving Cube Roots

Section 9.7 – Objective 1: Evaluate Functions Involving Radicals
Video Length – 2:48

We are now going to look at functions involving radicals.

1. **Example:** For the function $f(x) = \sqrt{-3x-5}$, find $f(-3)$ and $f(-11)$.

Final answer: _____

Course:
Instructor:
Name:
Section:

Section 9.7 – Objective 2: Find the Domain of a Function Involving a Radical
Video Length – 7:06

2. **Example:** Find the domain of $f(x) = \sqrt{-3x - 5}$.

   In general if the index on the radical is _____, _____
   _____.

   If the index on the radical is _____, _____.

   Set-builder notation: _____

   Interval notation: _____

3. **Example:** Find the domain of $f(x) = \sqrt[3]{2x + 1}$.

   Set-builder notation: _____

   Interval notation: _____

Course:
Instructor:
Name:
Section:

Section 9.7 – Objective 3: Graph Functions Involving Square Roots
Video Length – 6:31

4. **Example:** For the function $f(x) = \sqrt{x+4}$,

   (a) Find the domain.

   Set-builder notation: _____

   Interval notation: _____

   (b) Graph the function.

   (c) Determine the range from the graph.

   Set-builder notation: _____

   Interval notation: _____

344

Course: Name:
Instructor: Section:

Section 9.7 – Objective 4: Graph Functions Involving Cube Roots
Video Length – 6:29

5. **Example:** For the function $f(x) = \sqrt[3]{x} + 1$,

   (d) Find the domain.

   Set-builder notation: _____

   Interval notation: _____

   (e) Graph the function.

   (f) Determine the range from the graph.

   Set-builder notation: _____

   Interval notation: _____

Course: Name:
Instructor: Section:

Section 9.8 Video Guide
Radical Equations and Their Applications

Objectives:
1. Solve Radical Equations Containing One Radical
2. Solve Radical Equations Containing Two Radicals
3. Solve for a Variable in a Radical Equation

Section 9.8 – Objective 1: Solve Radical Equations Containing One Radical
Part I – Text Example 1
Video Length – 4:13

Definition
When the variable in an equation occurs in a radicand, the equation is called a _____ _____ .

1. **Example:** Solve and check: $\sqrt{5x-1} = 3$

      Final answer: _____

Course: Name:
Instructor: Section:

Section 9.8 – Objective 1: Solve Radical Equations Containing One Radical
Part II – Text Example 3
Video Length – 6:35

2. **Example:** Solve: $y = \sqrt{2y+9} - 3$

   Final answer: _____

It is ABSOLUTELY, POSITIVELY IMPORTANT to CHECK YOUR ANSWERS!!!

Course: Name:
Instructor: Section:

Section 9.8 – Objective 2: Solve Radical Equations Containing Two Radicals
Video Length – 10:59

We are now going to solve radical equations involving two radicals.

Note: Take a deep breath...Ready?

3. **Example:** Solve: $\sqrt{3y+1} - \sqrt{y-4} = 3$

Final answer: _____

Course: Name:
Instructor: Section:

Section 9.8 – Objective 3: Solve for a Variable in a Radical Equation
Video Length – 1:48

4. **Example:** Solve $r = \sqrt[3]{\dfrac{3V}{4\pi}}$ for V.

Final answer: _____

Course:
Instructor:
Name:
Section:

Section 9.9 Video Guide
The Complex Number System

Objectives:
1. Evaluate the Square Root of Negative Real Numbers
2. Add or Subtract Complex Numbers
3. Multiply Complex Numbers
4. Divide Complex Numbers
5. Evaluate the Powers of i

Section 9.9 – The Complex Number System
Video Length – 9:29

Nonnegativity Property of Real Numbers
For any real number a, $a^2 \geq 0$.

Because the square of any real number is never negative, there is no real number x for which

$$x^2 = -1.$$

Definition
The _____ _____, denoted by _____, is the number whose square is −1. That is

____ = ____

Definition
_____ _____ are numbers of the form _____, where a and b are real numbers. The real number a is called the _____ _____ of the number $a+bi$; the real number b is called the _____ _____ of $a+bi$.

$$3 + 5i$$
_____ _____

Definition
When a complex number is written in the form $a+bi$, where a and b are real numbers, it is written in _____ _____. Any number of the form bi is called a _____ _____ _____.

Course: Name:
Instructor: Section:

Complex Numbers That Are Not Real	**Pure Imaginary Numbers**
Rational numbers	**Irrational numbers**

Course: Name:
Instructor: Section:

Section 9.9 – Objective 1: Evaluate the Square Root of Negative Real Numbers
Video Length – 6:42

Now what we are going to do is evaluate square roots of negative numbers.

1.   **Example:** Simplify:

    (a) $\sqrt{-49}$ (a) **Final answer:** $\sqrt{-49}$ = _____

    (b) $\sqrt{-27}$ (b) **Final answer:** $\sqrt{-27}$ = _____

    (c) $\sqrt{-125}$ (c) **Final answer:** $\sqrt{-125}$ = _____

2.   **Example:** Write each of the following in standard form:

    (a) $4 - \sqrt{-36}$ (a) **Final answer:** $4 - \sqrt{-36}$ = _____

    (b) $-5 + \sqrt{-32}$ (b) **Final answer:** $-5 + \sqrt{-32}$ = _____

    (c) $3 - \sqrt{-25}$ (c) **Final answer:** $3 - \sqrt{-25}$ = _____

Course: Name:
Instructor: Section:

Section 9.9 – Objective 2: Add or Subtract Complex Numbers
Video Length – 5:28

Sum of Complex Numbers

_____ + _____ = _____ + _____

3. **Example:** Add each of the following:

   (a) $(4-7i)+(6+3i)$ (a) **Final answer:** $(4-7i)+(6+3i)=$ _____

   (b) $(-7-8i)+(2+10i)$ (b) **Final answer:** $(-7-8i)+(2+10i)=$ _____

Difference of Complex Numbers

_____ − _____ = _____ + _____

4. **Example:** Subtract each of the following:

   (a) $(6-9i)-(3+6i)$ (a) **Final answer:** $(6-9i)-(3+6i)=$ _____

   (b) $(-4+8i)-(2+7i)$ (b) **Final answer:** $(-4+8i)-(2+7i)=$ _____

5. **Example:** Add: $\left(5+\sqrt{-36}\right)+\left(2-\sqrt{-49}\right)$

   Final answer: $\left(5+\sqrt{-36}\right)+\left(2-\sqrt{-49}\right)=$ _____

Course:
Instructor:
Name:
Section:

Section 9.9 – Objective 3: Multiply Complex Numbers
Part I – Text Example 5
Video Length – 6:18

We are now going to multiply complex numbers (*yaaay!!!*)

6. **Example:** Multiply the following:

   (a) $9i(4+5i)$

   Final answer: $9i(4+5i) = $ _____

   (b) $(6-2i)(3+i)$

   Final answer: $(6-2i)(3+i) = $ _____

   (c) $(-4+8i)(2+7i)$

   Final answer: $(-4+8i)(2+7i) = $ _____

Course: Name:
Instructor: Section:

Section 9.9 – Objective 3: Multiply Complex Numbers
Part II – Text Example 7
Video Length – 4:11

Complex Conjugate
If $a+bi$ is a complex number, then its _____ is defined as _____ .

7. **Example:** Find the product of $6-9i$ and its conjugate.

    Final answer: _____

In general

$$(a+bi)(a-bi)$$

When you multiply a complex number and its conjugate you are always going to get a _____ _____ .

Course: Name:
Instructor: Section:

Section 9.9 – Objective 4: Divide Complex Numbers
Video Length – 7:58

8. **Example:** Divide the following: $\dfrac{6-i}{4+3i}$

   Final answer: $\dfrac{6-i}{4+3i} = $ _____

9. **Example:** Divide the following: $\dfrac{-4+5i}{6i}$

   Final answer: $\dfrac{-4+5i}{6i} = $ _____

Course: Name:
Instructor: Section:

Section 9.9 – Objective 5: Evaluate Powers of i
Video Length – 5:53

The **powers of i** follow a pattern.

i^1 i^5 i^9

i^2 i^6 i^{10}

i^3 i^7 i^{11}

i^4 i^8

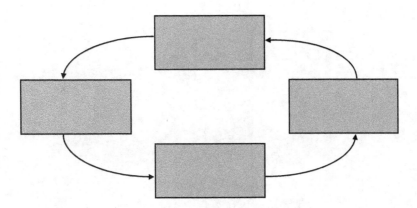

10. Example: Evaluate the following:

(a) i^{35}

Final answer: $i^{35} =$ _____

(b) i^{-26}

Final answer: $i^{-26} =$ _____

Course:
Instructor:

Name:
Section:

Course: Name:
Instructor: Section:

Section 10.1 Video Guide
Solving Quadratic Equations by Completing the Square

Objectives:
1. Solve Quadratic Equations Using the Square Root Property
2. Complete the Square in One Variable
3. Solve Quadratic Equations by Completing the Square
4. Solve Problems Using the Pythagorean Theorem

Section 10.1 – Objective 1: Solve Quadratic Equations Using the Square Root Property
Video Length – 12:12

Recall that a quadratic equation is an equation of the form_____ , _____.

Square Root Property
If $x^2 = p$, then $x =$ ____ or $x =$ _____ . (*Note:* $p \geq 0$)

1. **Example:** Solve the equation: $x^2 = 36$

    Final answer: _____

Course: Name:
Instructor: Section:

2. **Example:** Solve the equation: $3x^2 = 150$

Write the steps in words	Show the steps with math
Step 1	
Step 2	
Step 3	
Step 4	

Final answer: _____

3. **Example:** Solve: $(2x+3)^2 = 10$

Final answer: _____

Course: Name:
Instructor: Section:

Section 10.1 – Objective 2: Complete the Square in One Variable
Video Length – 6:09

The idea behind _____ ____ _____ is to "adjust" the left side of a quadratic

equation of the form $x^2 + bx + c = 0$ in order to make it a perfect square trinomial.

Do you remember what a perfect square trinomial looks like?

Obtaining a Perfect Square Trinomial
Step 1: Identify the coefficient of the _____-_____ term.

Step 2: _____ this coefficient by _____ and then _____ the result.

Step 3: _____ this result to both sides of the equation.

Consider

$$n^2 + 10n$$

Let's try

$$x^2 - 16x$$

Now try

$$z^2 + 7z$$

Course: Name:
Instructor: Section:

Section 10.1 – Objective 3: Solve Quadratic Equations by Completing the Square
Part I – Text Example 6
Video Length – 5:10

So let's solve a quadratic equation by completing the square.

4. **Example:** Solve: $x^2 - 6x - 7 = 0$

Write the steps in words	Show the steps with math
Step 1	
Step 2	
Step 3	
Step 4	
Step 5 Note: CHECK YOUR WORK!!!	

Final answer: _____

Note: You can also solve the quadratic equation $x^2 - 6x - 7 = 0$ by factoring. Try it. Did you get the same answer?

Course:
Instructor:

Name:
Section:

Section 10.1 – Objective 3: Solve Quadratic Equations by Completing the Square
Part II – Text Example 7
Video Length – 7:36

5. **Example:** Solve: $2x^2 + 5x - 3 = 0$

Final answer: _____

Course:
Instructor:
Name:
Section:

Section 10.1 – Objective 4: Solve Problems Using the Pythagorean Theorem
Video Length – 4:59

The Pythagorean Theorem is a statement regarding right triangles.

Pythagorean Theorem
In a right triangle, the square of the length of the hypotenuse is equal to the sum of the squares of the lengths of the legs.

____ + ____ = ____

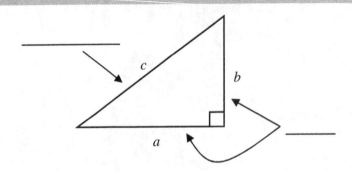

6. **Example:** A baseball diamond is square. Each side of the square is 90 feet long. How far is it from home plate to second base?

Draw diagram here:

Final answer: _____

Section 10.2 Video Guide
Solving Quadratic Equations by the Quadratic Formula

Objectives:
1. Solve Quadratic Equations Using the Quadratic Formula
2. Use the Discriminant to Determine the Nature of Solutions of a Quadratic Equation
3. Model and Solve Problems Involving Quadratic Equations

Section 10.2 – Objective 1: Solve Quadratic Equations Using the Quadratic Formula
Part I
Video Length – 7:29

Proof of the Quadratic Formula:

Course:
Instructor:
Name:
Section:

Section 10.2 – Objective 1: Solve Quadratic Equations Using the Quadratic Formula
Part II – Text Example 1
Video Length – 5:50

Quadratic Formula
The solution(s) to the quadratic equation $ax^2 + bx + c = 0$, $a \neq 0$, are given by the **quadratic formula**

$$x = \underline{\qquad\qquad}.$$

Note: Each and every time you do a problem utilizing the quadratic formula, you should write it down. This way, you will ingrain the formula into your memory.

1. **Example:** Solve: $2x^2 + 11x = -15$

Write the steps in words	Show the steps with math
Step 1	
Step 2	
Step 3	
Step 4	

Final answer: _____

Course: Name:
Instructor: Section:

Section 10.2 – Objective 1: Solve Quadratic Equations Using the Quadratic Formula
Part III – Text Example 2
Video Length – 5:38

2. **Example:** Solve: $y^2 - 2 = -4y$

Final answer: _____

Course:
Instructor:
Name:
Section:

Section 10.2 – Objective 1: Solve Quadratic Equations Using the Quadratic Formula
Part IV – Text Example 4
Video Length – 8:13

3. **Example:** Solve: $3m^2 + 2m + 1 = 0$

Final answer: _____

Course: Name:
Instructor: Section:

Section 10.2 – Objective 2: Use the Discriminant to Determine the Nature of Solutions of a Quadratic Equation
Part I
Video Length – 11:05

Definition

In the quadratic formula $x = $ ——————, the quantity _____ is called the

_____ of the quadratic equation, because its value tells us the _____ of

solutions and the _____ of solution.

Discriminant	Number of Solutions	Type of Solutions

4. **Example:** Use the discriminant to determine the type of solutions of the equation.

(a) $4x^2 - 20x + 25 = 0$

Final answer: _____

(b) $x^2 - 3(x-8) = 2x$

Final answer: _____

Course: Name:
Instructor: Section:

Section 10.2 – Objective 2: Use the Discriminant to Determine the Nature of Solutions of a Quadratic Equation
Part II – Text Example 5
Video Length – 11:41

5. **Example:** Use the discriminant to determine the type of solutions of the equation.

   (a) $4x^2 - 20x + 25 = 0$

   Final answer: _____

   (b) $x^2 - 3(x-8) = 2x$

   Final answer: _____

   (c) **Example:** Solve: $3x^2 - 4x + 2 = 0$

   Final answer: _____

Course: Name:
Instructor: Section:

Section 10.2 – Objective 3: Model and Solve Problems Involving Quadratic Equations
Video Length – 7:38

6. **Example:** The revenue R received by a company selling x pairs of sunglasses per week is given by the function $R(x) = -0.1x^2 + 70x$.

(a) Find and interpret the values of $R(17)$ and $R(25)$.

   Final answer: _____

(b) How many pairs of sunglasses must be sold in order for revenue to be $10,000 per week?

   Final answer: _____

(c) How many pairs of sunglasses must be sold in order for revenue to be $12,250 per week?

   Final answer: _____

Course: Name:
Instructor: Section:

Section 10.3 Video Guide
Solving Equations Quadratic in Form

Objectives:
1. Solve Equations That are Quadratic in Form

Section 10.3 – Objective 1: Solve Equations That are Quadratic in Form
Part I – Text Example 1
Video Length – 9:04

Definition
If a substitution u transforms an equation into one of the form

then the original equation is called an _____ _____ ____ _____.

Examples:

1. **Example:** Solve: $x^4 - 11x^2 + 18 = 0$

Final answer: _____

Course: Name:
Instructor: Section:

Section 10.3 – Objective 1: Solve Equations That are Quadratic in Form
Part II – Text Example 5
Video Length – 8:19

Note: This is a really nice problem. It shows that it really pays to ALWAYS CHECK your answers.

2. **Example:** Solve: $2x^{1/2} - x^{1/4} - 1 = 0$

Final answer: _____

Course: Name:
Instructor: Section:

Section 10.4 Video Guide
Graphing Quadratic Functions Using Transformations

Objectives:
1. Graph Quadratic Functions of the Form $f(x) = x^2 + k$
2. Graph Quadratic Functions of the Form $f(x) = (x-h)^2$
3. Graph Quadratic Functions of the Form $f(x) = ax^2$
4. Graph Quadratic Functions of the Form $f(x) = ax^2 + bx + c$
5. Find a Quadratic Function from Its Graph

Section 10.4 – Objective 1: Graph Quadratic Functions of the Form $f(x) = x^2 + k$
Video Length – 13:30

Definition
A _____ _____ is a function of the form

where a, b, and c are real numbers and $a \neq 0$.

_____ can be used to graph a quadratic function.

1. **Example:** Graph the functions on one coordinate plane.

$f(x) = x^2$ $g(x) = x^2 + 3$

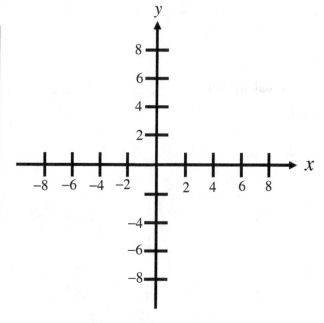

Notice that the graph of $g(x)$ is the graph of $f(x)$ shifted _____ .

2. **Example:** Graph the functions on one coordinate plane.

$f(x) = x^2$

x	f(x)

$g(x) = x^2 - 2$

x	g(x)

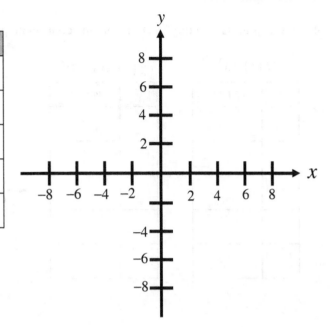

Notice that the graph of $g(x)$ is the graph of $f(x)$ shifted _____.

3. **Example:** Graph the function $h(x) = x^2 + 4$.

Course: _____ Name: _____
Instructor: _____ Section: _____

Section 10.4 – Objective 2: Graph Quadratic Functions of the Form $f(x)=(x-h)^2$

Video Length – 11:30

4. **Example:** Graph the functions on one coordinate plane.

$f(x)=x^2$

x	$f(x)$

$g(x)=(x+3)^2$

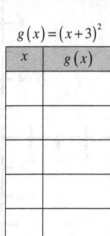

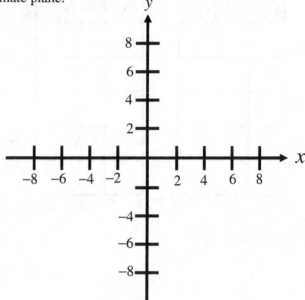

Notice that the graph of $g(x)$ is the graph of $f(x)$ shifted _____.

5. **Example:** Graph the functions on one coordinate plane.

$f(x)=x^2$

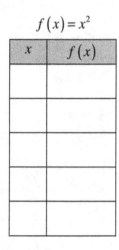

$g(x)=(x-1)^2$

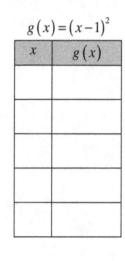

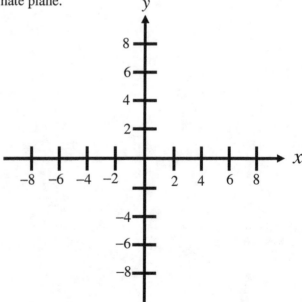

Notice that the graph of $g(x)$ is the graph of $f(x)$ shifted _____.

Course: Name:
Instructor: Section:

6. **Example:** Graph the function $f(x)=(x-3)^2$.

Course: Name:
Instructor: Section:

Section 10.4 – Objective 3: Graph Quadratic Functions of the Form $f(x)=ax^2$

Video Length – 14:12

7. **Example:** Graph the functions on one coordinate plane.

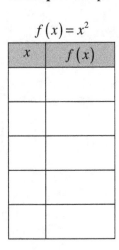  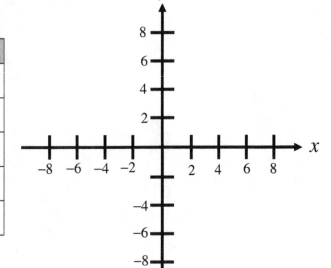

The graph is _____ _____ by a factor of _____.

8. **Example:** Graph the functions on one coordinate plane.

 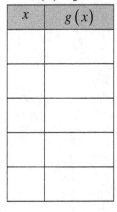 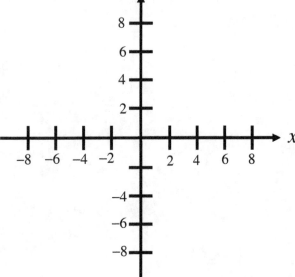

The graph is _____ _____ (*by a factor of* $\frac{1}{2}$).

Course: Name:
Instructor: Section:

9. **Example:** Graph the functions on one coordinate plane.

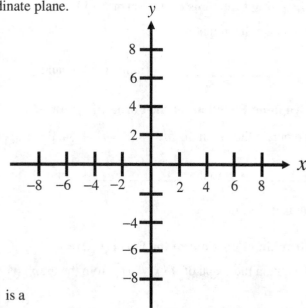

Notice that the graph of $g(x)$ is a

_____ of the graph of

$f(x)$ _____ .

Properties of the Form $f(x) = ax^2$

- If $a > 0$, the graph of $f(x) = ax^2$ will open _____ . In addition, if $0 < a < 1$, the opening in the graph will be "_____" than that of $y = x^2$. If $a > 1$, the opening in the graph will be "_____" than that of $y = x^2$.

- If $a < 0$, the graph of $f(x) = ax^2$ will open _____ . In addition, if $0 < |a| < 1$, the opening in the graph will be "_____" than that of $y = x^2$. If $|a| > 1$, the opening in the graph will be "_____" than that of $y = x^2$.

- When $|a| > 1$, we say the graph is _____ _____ by a factor of $|a|$.

  When $0 < |a| < 1$, we say that the graph is _____ _____ by a factor of $|a|$.

Course: Name:
Instructor: Section:

Graphing Using Transformations

Graphing Functions of the Form $f(x) = x^2 + k$

To obtain the graph of $f(x) = x^2 + k$ from the graph of $y = x^2$, shift the graph of $y = x^2$

_____ _____ _____ units if $k > 0$ and _____ _____ _____ units if $k < 0$.

Graphing Functions of the Form $f(x) = (x-h)^2$

To obtain the graph of $f(x) = (x-h)^2$ from the graph of $y = x^2$, shift the graph of $y = x^2$

_____ to the _____ _____ units if $h > 0$ and _____ to the _____ _____ units if $h < 0$.

Graphing Functions of the Form $f(x) = ax^2$

To obtain the graph of $f(x) = ax^2$ from the graph of $y = x^2$, _____ each _____ on the graph of $y = x^2$ by _____ .

Course: Name:
Instructor: Section:

Section 10.4 – Objective 4: Graph Quadratic Functions of the Form $f(x)=ax^2+bx+c$
Part I – Text Example 7
Video Length – 13:28

Definition
The graph of a quadratic function is a _____.

The _____ is the lowest or highest point of a parabola.

The _____ ____ _____ is the vertical line passing through the vertex.

Graphing Quadratic Functions Using Transformations

Step 1: Write the function $f(x)=ax^2+bx+c$ as _____ by completing the square in x.

Step 2: Graph the function _____ using transformations.

10. Example: Graph $f(x)=x^2+4x-1$ using transformations.

Course:
Instructor:
Name:
Section:

Section 10.4 – Objective 4: Graph Quadratic Functions of the Form $f(x) = ax^2 + bx + c$
Part II – Text Example 8
Video Length – 19:53

11. Example: Graph $f(x) = -3x^2 - 6x - 1$ using transformations.

Course: Name:
Instructor: Section:

Section 10.4 – Objective 5: Find a Quadratic Function from Its Graph
Video Length – 6:33

If we are given the vertex (h,k), and one additional point on the graph of a quadratic function, we can find the quadratic function $f(x) = ax^2 + bx + c$ that results in the given graph.

12. **Example:** Find the quadratic function whose graph has a vertex of $(-2,5)$ and passes through the point $(0,-7)$.

   Final answer: _____

Course: Name:
Instructor: Section:

Section 10.5 Video Guide
Graphing Quadratic Functions Using Properties

Objectives:
1. Graph Quadratic Functions of the Form $f(x) = ax^2 + bx + c$
2. Find the Maximum or Minimum Value of a Quadratic Function
3. Model and Solve Optimization Problems Involving Quadratic Functions

Section 10.5 – Objective 1: Graph Quadratic Functions of the Form $f(x) = ax^2 + bx + c$

Part I – Text Example 1
Video Length – 21:08

Consider the quadratic function

$$f(x) = ax^2 + bx + c \qquad , a \neq 0$$

Properties of the Graph of a Quadratic Function

$$f(x) = ax^2 + bx + c, \quad a \neq 0$$

Vertex =   Axis of symmetry; the line $x =$

Parabola opens _____ if _____; the vertex is a _____ point.

Parabola opens _____ if _____; the vertex is a _____ point.

Consider the function $f(x) = 3x^2 + 12x - 5$.

Course: Name:
Instructor: Section:

The *x*-intercepts of a Quadratic Function

1. If the discriminant _____, the graph of $f(x) = ax^2 + bx + c$ has _____ distinct *x*-intercepts and so will cross the *x*-axis in _____ places.

2. If the discriminant _____, the graph of $f(x) = ax^2 + bx + c$ has _____ *x*-intercept and touches the *x*-axis at its _____.

3. If the discriminant _____, the graph of $f(x) = ax^2 + bx + c$ has _____ *x*-intercept and so will not cross or touch the *x*-axis.

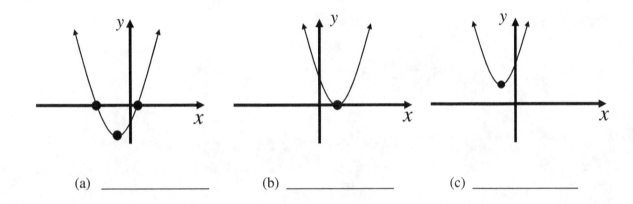

(a) _____   (b) _____   (c) _____

_____   _____   _____

1. **Example:** Graph $f(x) = 3x^2 + 12x - 5$ using its properties.

Course:
Instructor:

Name:
Section:

Section 10.5 – Objective 1: Graph Quadratic Functions of the Form $f(x) = ax^2 + bx + c$
Part II – Text Example 4
Video Length – 5:36

2. **Example:** Graph $f(x) = -x^2 + 4x - 7$ using its properties.

Course: Name:
Instructor: Section:

Section 10.5 – Objective 2: Find the Maximum or Minimum Value of a Quadratic Function
Video Length – 5:28

The graph of a quadratic function has a vertex at _____ .

Definition

The vertex will be the _____ point on the graph if _____ and $f\left(-\dfrac{b}{2a}\right)$ will be the _____ _____ of f.

The vertex will be the _____ point on the graph if _____ and $f\left(-\dfrac{b}{2a}\right)$ will be the _____ _____ of f.

Opens _____
_____

Opens _____
_____

3. **Example:** Determine whether the quadratic function $f(x) = -3x^2 + 12x - 1$ has a maximum or minimum value. Find the value.

Final answer: _____

Course:
Instructor:
Name:
Section:

Section 10.5 – Objective 3: Model and Solve Optimization Problems Involving Quadratic Functions

Video Length – 5:58

4. **Example:** The Great Lakes Tour Company offers one-day tours at the rate of $90 per person for each of the first 30 people. For larger groups, each person receives a $0.50 discount. How many people will be required for the tour company to maximize revenue? What is the maximum revenue?

Final answer: _____

Course: Name:
Instructor: Section:

Section 10.6 Video Guide
Polynomial Inequalities

Objectives:
1. Solve Quadratic Inequalities
2. Solve Polynomial Inequalities

Section 10.6 – Objective 1: Solve Quadratic Inequalities
Part I – Text Example 1
Video Length – 13:43

Definition
A _____ _____ is an inequality of the form

_____ or _____

or

_____ or _____

where $a \neq 0$.

We will go through two different approaches for solving quadratic inequalities. This first approach is a graphical approach and the second is an algebraic approach.

Note: The work for the following example takes up two slides. So make sure you save some room for the graph on the second slide.

1. **Example:** Solve the inequality using the graphical method: $x^2 + 3x - 28 \geq 0$.

Set-builder: _____   Interval notation: _____

Course: Name:
Instructor: Section:

Section 10.6 – Objective 1: Solve Quadratic Inequalities
Part II – Text Example 2
Video Length – 9:13

2. **Example:** Solve the inequality: $x^2 + 3x - 28 \geq 0$.

Write the steps in words	Show the steps with math
Step 1	
Step 2	
Step 3	
Step 4	

Set-builder: _____

Interval notation: _____

Course:
Instructor:
Name:
Section:

Section 10.6 – Objective 1: Solve Quadratic Inequalities
Part III – Text Example 3
Video Length – 10:30

3. **Example:** Solve the inequality: $-x^2 + 8 > 2x$

   Solve graphically:

   Solve algebraically:

   Set-builder: _____

   Interval notation: _____

Course: Name:
Instructor: Section:

Section 10.6 – Objective 2: Solve Polynomial Inequalities
Video Length – 5:24

Remember, the zeros of a function f are the values that cause the function's value to be 0. If r is a zero, then $f(r)=0$. So, the zeros of a function are the same as the x-intercepts of the graph of the function.

4. **Example:** Solve the inequality: $(x+3)^2(x-1)(x-4) \leq 0$.

Set-builder: _____

Course: Name:
Instructor: Section:

Section 10.7 Video Guide
Rational Inequalities

Objectives:
1. Solve a Rational Inequality

Section 10.7 – Objective 1: Solve a Rational Inequality
Part I – Text Example 1
Video Length – 15:37

Definition
A _____ _____ is an inequality that contains a rational expression.

Examples:

There are two keys to solving rational inequalities:

1. The quotient of two positive numbers is positive ; the quotient of a positive and negative number is negative ; and the quotient of two negative numbers is positive.

2. A rational expression may change signs on either side of the value of the makes the rational expression equal _____ or for values for which the rational expression is _____.

1. **Example:** Solve: $\dfrac{x+8}{x+2} \leq 0$

Write the steps in words	Show the steps with math
Step 1	
Step 2	
Step 3	
Step 4	

Set-builder: _____

Interval notation: _____ Graph: _____

Course: Name:
Instructor: Section:

Section 10.7 – Objective 1: Solve a Rational Inequality
Part II – Text Example 2
Video Length – 6:31

2. Example: $\dfrac{2x+3}{x-2} < 1$

Set-builder: _____

Interval notation: _____

Graph: _____→

Course: Name:
Instructor: Section:

Section 11.1 Video Guide
Composite Functions and Inverse Functions

Objectives:
1. Form the Composite Function
2. Determine Whether a Function is One-to-One
3. Find the Inverse of a Function Defined by a Map or Set of Ordered Pairs
4. Obtain the Graph of the Inverse Function from the Graph of the Function
5. Find the Inverse of a Function Defined by an Equation

Section 11.1 – Objective 1: Form the Composite Function
Part I – Text Example 1
Video Length – 10:18

Suppose you have a job that pays $10/hr.

The idea of using a function's output as the input to another function is called **composition**.

Definition
Given two functions f and g, the _____ _____, denoted by _____ (read as "f composed with g"), is defined by _____ = _____ .

In other words, the output of the function g becomes the input of the function f.

1. **Example:** Suppose that $f(x) = 2x^2 + 3$ and $g(x) = 4x^3 + 1$. Find:

    (a) $(f \circ g)(1)$ (b) $(g \circ f)(1)$ (c) $(f \circ f)(-2)$ (d) $(g \circ g)(-1)$

       Final answer: _____   **Final answer:** _____   **Final answer:** _____   **Final answer:** _____

Course: Name:
Instructor: Section:

Section 11.1 – Objective 1: Form the Composite Function
Part II – Text Example 2
Video Length – 3:48

Rather than evaluating a composite function at a specific value, we can also form a composite function that is written in terms of the independent variable, x.

2. **Example:** Suppose that $f(x) = \dfrac{1}{x+4}$ and $g(x) = \dfrac{4}{x-2}$. Find $(f \circ g)(x)$.

Final answer: _____

Course: _____ Name: _____
Instructor: _____ Section: _____

Section 11.1 – Objective 2: Determine Whether a Function is One-to-One
Part I – Text Example 3
Video Length – 8:47

Recall that a function is a relation in which each input of the relation corresponds to exactly one output of the relation.

To put it another way, we do not have a function when _____

_____

Is the following relation a function?

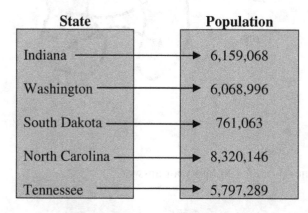

Is this relation a function?

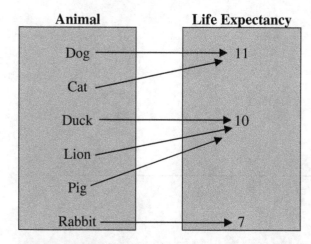

Definition
A function is _____-____-_____ if any two different inputs in the domain correspond to two different outputs in the range. That is, if x_1 and x_2 are two different inputs of a function f, then $f(x_1) \neq f(x_2)$.

Which of the above functions is one-to-one?

Course: Name:
Instructor: Section:

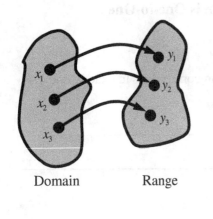

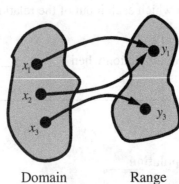

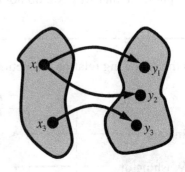

3. **Example:** Determine whether the function is one-to-one? Explain your answer?

(a)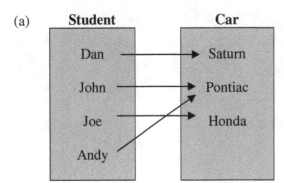

    Final answer: _____

(b) $\{(1,5),(2,8),(3,11),(4,14)\}$

    Final answer: _____

Course: Name:
Instructor: Section:

Section 11.1 – Objective 2: Determine Whether a Function is One-to-One
Part II – Text Example 4
Video Length – 3:47

Horizontal-line Test
If _____ _____ _____ intersects the graph of a function f in at most _____ point, then f is one-to-one.

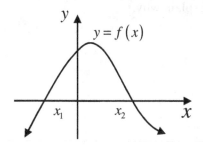

4. **Example:** Use the graph to determine whether its corresponding function is one-to-one.

   (a)

   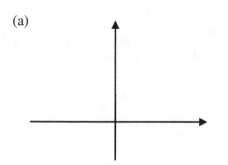

   Final answer: _____

   (b)

   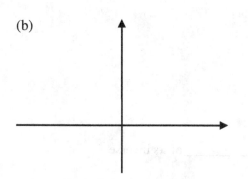

   Final answer: _____

Course:
Instructor:
Name:
Section:

Section 11.1 – Objective 3: Find the Inverse of a Function Defined by a Map or Set of Ordered Pairs
Video Length – 5:11

Note: There is a lot going on in this problem. Listen carefully to his explanations of why a function has an inverse and why one does not. Also, pay close attention to the relationship between the domain and range of a function to the range and domain of its inverse.

5. **Example:** Determine which of the following functions is one-to-one? If the function is one-to-one, find its inverse. If the function is not one-to-one, explain why.

(a)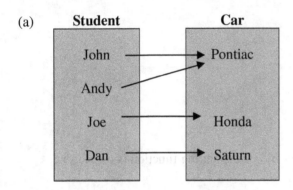

(b)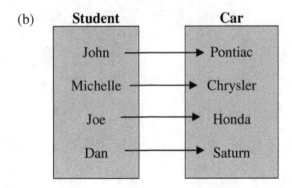

(c) $\{(1,5),(2,8),(3,11),(4,14)\}$

The _____ of the function is the same as the _____ of its inverse.

The _____ of the inverse is the same as the _____ of the original function.

400 Copyright © 2014 Pearson Education, Inc.

Course: Name:
Instructor: Section:

Section 11.1 – Objective 4: Obtain the Graph of the Inverse Function from the Graph of the Function
Video Length – 4:45

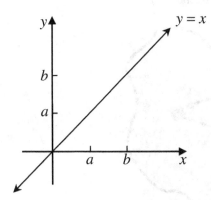

Suppose (a,b) is a point on the graph of some function f. What point would be on the graph of its inverse? _____ . It turns out that the line segment joining (a,b) and (b,a) is _____ to the line $y = x$. We know this because the slope of the line joining (a,b) and (b,a) is _____ .

Additionally, since the distance from (b,a) to $y = x$ and (a,b) to $y = x$ is the same, we can conclude that the graph of a function f and its inverse are going to be _____ about the line _____ .

Theorem
The graph of a function f and the graph of its inverse f^{-1} are _____ with respect to the line _____ .

6. **Example:** The graph of a one-to-one function f is given. Draw the graph of the inverse function f^{-1}.

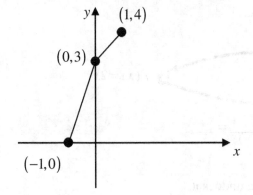

Section 11.1 – Objective 5: Find the Inverse of a Function Defined by an Equation
Video Length – 13:36

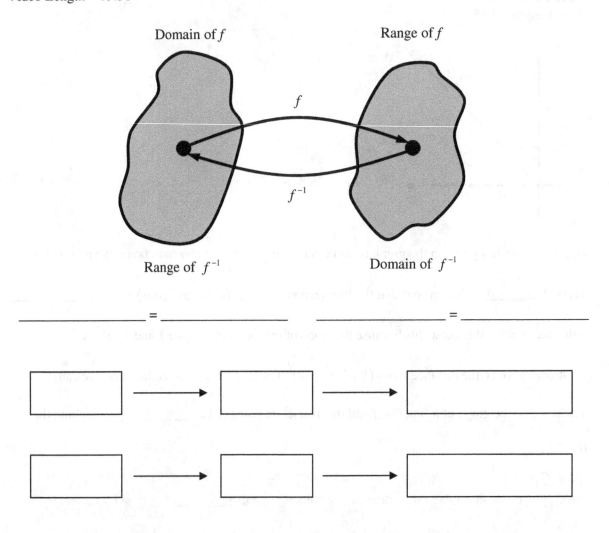

_____ = _____   _____ = _____

_____ = _____ where x is in the domain of _____

_____ = _____ where x is in the domain of _____

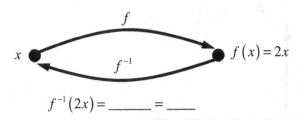

$f^{-1}(2x) = $ _____ = _____

Remember, the function does something and the inverse undoes it.

Course: Name:
Instructor: Section:

7. **Example:** Verify that the functions $g(x) = x^3 + 2$ and $g^{-1}(x) = \sqrt[3]{x-2}$ are inverses of each other.

8. **Example:** Find the inverse function of the one-to-one function $f(x) = \dfrac{2x-1}{x+1}$.

Write the steps in words	Show the steps with math
Step 1	
Step 2	
Step 3	

Final answer: _____

Course: Name:
Instructor: Section:

Section 11.2 Video Guide
Exponential Functions

Objectives:
1. Evaluate Exponential Expressions
2. Graph Exponential Functions
3. Define the Number e
4. Solve Exponential Equations
5. Use Exponential Models That Describes Our World

Section 11.2 – Objective 1: Evaluate Exponential Expressions
Video Length – 11:13

When you were first introduced to exponents, you were taught how to raise a real number to a positive integer power. You were then taught how to raise a nonzero real number to the zero power. Then you were taught how to raise real numbers to integer powers. Finally, you were taught how to raise a real number to a rational power. Now we are going to talk about raising a real number to a real power.

1. **Example:** Using a calculator, evaluate:

    (a) $2^{1.4}$ (a) **Final answer:** $2^{1.4} \approx$ _____

    (b) $2^{1.41}$ (b) **Final answer:** $2^{1.41} \approx$ _____

    (c) $2^{1.414}$ (c) **Final answer:** $2^{1.414} \approx$ _____

    (d) $2^{1.4142}$ (d) **Final answer:** $2^{1.4142} \approx$ _____

    (e) $2^{\sqrt{2}}$ (e) **Final answer:** $2^{\sqrt{2}} \approx$ _____

Law of Exponents
If s, t, a, and b are real numbers with $a > 0$ and $b > 0$, then

_____ = _____   _____ = _____   _____ = _____

_____ = _____   _____ = _____ = _____   _____ = _____

An exponential function is a function of the form $f(x) =$ _____ , $c \neq 0$ and $a > 0$ are

real numbers. The real number c is called the _____ _____ .

Course: Name:
Instructor: Section:

Section 11.2 – Objective 2: Graph Exponential Functions
Part I – Text Example 2
Video Length – 4:32

2. **Example:** Graph the exponential function: $f(x) = 2^x$

The domain of the function is the set of all _____ numbers.

The range of the function is _____ or _____.

Properties of the Exponential Function $f(x) = a^x$, $a > 1$
1. The domain is the set of all real numbers; the range is the set of positive real numbers.
2. There are _____ x-intercepts; the y-intercept is _____.
3. The x-axis ($y = 0$) is a _____ asymptote as $x \to -\infty$.
4. $f(x) = a^x$, $a > 1$, is an _____ function and is _____-_____-_____.
5. The graph of f contains the points _____, _____, and _____.
6. The graph of f is _____ and _____, with no corners or gaps.

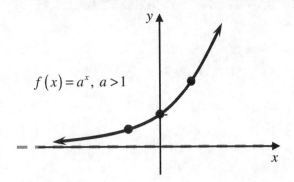

Copyright © 2014 Pearson Education, Inc.

Course:
Instructor:
Name:
Section:

Section 11.2 – Objective 2: Graph Exponential Functions
Part II – Text Example 3
Video Length – 2:49

3. **Example:** Graph the exponential function: $f(x) = \left(\dfrac{1}{2}\right)^x$

The domain of the function is the set of all _____ numbers.

The range of the function is _____ or _____ .

Properties of the Exponential Function $f(x) = a^x$, $0 < a < 1$
1. The domain is the set of all real numbers; the range is the set of positive real numbers.
2. There are _____ x-intercepts; the y-intercept is _____ .
3. The x-axis ($y = 0$) is a _____ asymptote as $x \to \infty$.
4. $f(x) = a^x$, $0 < a < 1$, is an _____ function and is ___-___-___ .
5. The graph of f contains the points _____, _____, and _____ .
6. The graph of f is _____ and _____ , with no corners or gaps.

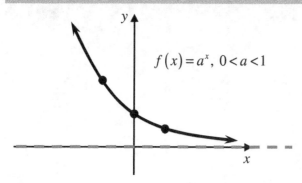

406   Copyright © 2014 Pearson Education, Inc.

Course:
Instructor:

Name:
Section:

Section 11.2 – Objective 3: Define the Number e
Video Length – 5:47

Definition
The **number e** is defined as the number that the expression

as $n \to \infty$.

In calculus, this is expressed using limit notation as

$$e = \lim_{n \to \infty} \left(1 + \frac{1}{n}\right)^n$$

Course: Name:
Instructor: Section:

Section 11.2 – Objective 4: Solve Exponential Equations
Video Length – 4:57

We are now going to solve exponential equations.

Property for Solving Exponential Equations

$$\text{If } \underline{\hspace{1cm}} = \underline{\hspace{1cm}}, \text{ then } \underline{\hspace{1cm}} = \underline{\hspace{1cm}}.$$

4. **Example:** Solve: $2^{3x-1} = 32$

   Final answer: _____

5. **Example:** Solve: $e^{2x-1} = \dfrac{1}{e^{3x}} \cdot \left(e^{-x}\right)^4$

   Final answer: _____

408 Copyright © 2014 Pearson Education, Inc.

Course: Name:
Instructor: Section:

Section 11.2 – Objective 5: Use Exponential Models That Describes Our World
Video Length – 2:04

6. **Example:** Between 12:00 PM and 1:00 PM, cars arrive at Citibank's drive-thru at the rate of 6 cars per hour (0.1 car per minute). The following formula from probability can be used to determine the probability that a car will arrive within *t* minutes of 12:00 PM:

$$F(t) = 1 - e^{-0.1t}$$

(a) Determine the probability that a care will arrive within 10 minutes of 12:00 PM (that is, before 12:10 PM).

Final answer: _____
Note: Write your answer in a complete sentence.

(b) Determine the probability that a care will arrive within 40 minutes of 12:00 PM (that is, before 12:40 PM).

Final answer: _____
Note: Write your answer in a complete sentence.

Course: Name:
Instructor: Section:

Section 11.3 Video Guide
Logarithmic Functions

Objectives:
1. Change Exponential Equations to Logarithmic Equations
2. Change Logarithmic Equations to Exponential Equations
3. Evaluate Logarithmic Functions
4. Determine the Domain of a Logarithmic Function
5. Graph Logarithmic Functions
6. Work with Natural and Common Logarithms
7. Solve Logarithmic Equations
8. Use Logarithmic Models That Describe Our World

Section 11.3 – Objective 1: Change Exponential Equations to Logarithmic Expressions
Video Length – 7:10

Definition
The _____ _____ to the _____ _____, where $a > 0$ and $a \neq 1$, is

denoted by _____ (read as "y is the logarithm to the base a of x") and is defined by

_____ if and only if _____ .

The domain of the logarithmic function $y = \log_a x$ is _____ .

1. **Example:** Change the exponential equation to an equivalent equation involving a logarithm.

   (a) $5^8 = t$ (a) **Final answer:** _____

   (b) $x^{-2} = 12$ (b) **Final answer:** _____

   (c) $4^y = 18$ (c) **Final answer:** _____

Course:
Instructor:

Name:
Section:

Section 11.3 – Objective 2: Change Logarithmic Equations to Exponential Expressions
Video Length – 1:59

2. **Example:** Write each logarithmic equation as an exponential equation.

   (a) $y = \log_2 21$

   (a) **Final answer:** _____

   (b) $\log_z 12 = 6$

   (b) **Final answer:** _____

Course: Name:
Instructor: Section:

Section 11.3 – Objective 3: Evaluate Logarithmic Functions
Video Length – 4:32

We will now find the exact value of a logarithmic expression.

3. **Example:** Find the exact value of each logarithmic expression.

_____ means _____

(a) $\log_3 81$

(a) **Final answer:** $\log_3 81 =$ _____

(b) $\log_2 \dfrac{1}{8}$

(b) **Final answer:** $\log_2 \dfrac{1}{8} =$ _____

Course: Name:
Instructor: Section:

Section 11.3 – Objective 4: Determine the Domain of a Logarithmic Function
Video Length – 8:52

Domain of the logarithmic function = _____ = _____

Range of the logarithmic function = _____ = _____

$$y = \log_a x \quad \text{(defining equation: } x = a^y\text{)}$$

   Domain: _____   Range: _____

4. **Example:** Find the domain of each logarithmic function.

 (a) $f(x) = \log_3(x-2)$ **(a) Final answer:** _____

 (b) $F(x) = \log_2\left(\dfrac{x+3}{x-1}\right)$ **(b) Final answer:** _____

Note: Listen to what he says about putting some thought into part (c). This idea will also reflect in part (d).

 (c) $h(x) = \log_2 |x-1|$ **(c) Final answer:** _____

 (d) $g(x) = \log_{\frac{1}{2}} x^2$ **(d) Final answer:** _____

Course:
Instructor:

Name:
Section:

Section 11.3 – Objective 5: Graph Logarithmic Functions
Video Length – 3:29

Note: Make sure to graph the corresponding logarithm functions.

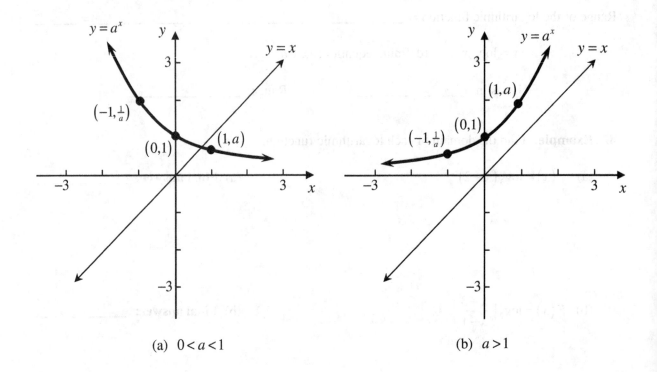

(a) $0 < a < 1$

(b) $a > 1$

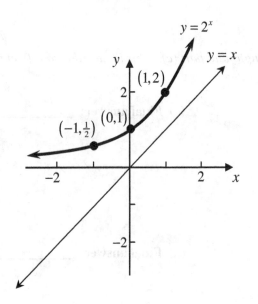

Course: Name:
Instructor: Section:

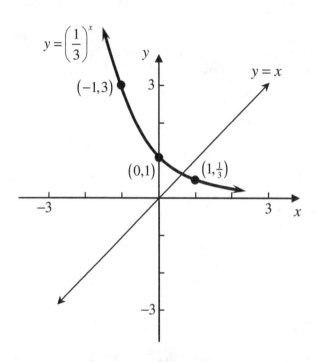

Properties of the Logarithmic Function $f(x) = \log_a x$

1. The domain is the set of positive real numbers; the range is the set of all real numbers.

2. The x-intercept of the graph is _____ . There is _____ y-intercept.

3. The y-axis ($x = 0$) is a _____ asymptote of the graph.

4. A logarithmic function is _____ if $0 < a < 1$ and _____ if $a > 1$.

5. The graph of f contains the points _____ , _____ , and _____ .

6. The graph of f is _____ and _____ , with no corners or gaps.

Course:
Instructor:
Name:
Section:

Section 11.3 – Objective 6: Work with Natural and Common Logarithms
Video Length – 4:02

Definition
The **natural logarithm**: _____ if and only if _____

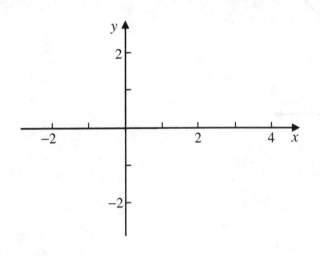

Definition
The **common logarithm**: _____ if and only if _____

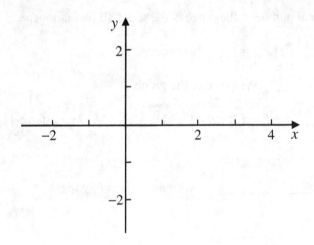

5. **Example:** Using a calculator, evaluate each of the following. Round answers to three decimal places.

    (a) $\log 25$ (a) **Final answer:** $\log 25 \approx$ _____

    (b) $\ln 25$ (b) **Final answer:** $\ln 25 \approx$ _____

    (c) $\ln 0.7$ (c) **Final answer:** $\ln 0.7 \approx$ _____

Course: Name:
Instructor: Section:

Section 11.3 – Objective 7: Solve Logarithmic Equations
Video Length – 3:38

We will now solve logarithmic equations.

6. **Example:** Solve:

   _____ means _____

   (a) $\log_2(2x+1) = 3$

   Final answer: _____

   (b) $\log_x 343 = 3$

   Final answer: _____

Course: Name:
Instructor: Section:

Section 11.3 – Objective 8: Using Logarithmic Models That Describe Our World
Video Length – 2:37

Common logarithms often are used when quantities vary from very large to very small. This is because the common logarithm can "scale down" the measurement.

Physicists define the **intensity of a sound wave** as the amount of energy the sound wave transmits through a given area.

> **Definition**
> The _____ L, measured in decibels, of a sound of intensity x, measured in watts per square meter, is

7. **Example:** A dripping faucet has an intensity level of 10^{-9} watt per square meter. How many decibels are there in the sound of a dripping faucet?

Final answer: _____

Note: Write your answer in a complete sentence.

Course: Name:
Instructor: Section:

Section 11.4 Video Guide
Properties of Logarithms

Objectives:
1. Understand the Properties of Logarithms
2. Write a Logarithmic Expression as a Sum or Difference of Logarithms
3. Write a Logarithmic Expression as a Single Logarithm
4. Evaluate a Logarithm Whose Base Is Neither 10 Nor e

Section 11.4 – Objective 1: Understand the Properties of Logarithms
Video Length – 7:04

1. **Example:** Deriving Properties of Logarithms

    (a) Show that $\log_a 1 = 0$.

    (b) Show that $\log_a a = 1$.

Properties of Logarithms
In the properties given next, M and a are positive real numbers, with $a \neq 1$, and r is any real number.

The number $\log_a M$ is the _____

_____.

_____ = _____

The logarithm to the base a of a raised to a power equals that power.

_____ = _____

2. **Example:** Find the exact value of each expression.

    (a) $3^{\log_3 18}$ (a) **Final answer:** $3^{\log_3 18} = $ _____

    (b) $2^{\log_2 (-5)}$ (b) **Final answer:** $2^{\log_2 (-5)} = $ _____

    (c) $\log_{\frac{1}{2}} \left(\frac{1}{2}\right)^{20}$ (c) **Final answer:** $\log_{\frac{1}{2}} \left(\frac{1}{2}\right)^{20} = $ ___

    (d) $\ln e^3$ (d) **Final answer:** $\ln e^3 = $ _____

Course:
Instructor:
Name:
Section:

Section 11.4 – Objective 2: Write a Logarithmic Expression as a Sum or Difference of Logarithms
Video Length – 10:50

Properties of Logarithms
In the following properties, M, N, and a are positive real numbers, with $a \neq 1$, and r is any real number.

The Log of a Product Equals the Sum of the Logs

$$\underline{\hspace{2cm}} = \underline{\hspace{3cm}}$$

The Log of a Quotient Equals the Difference of the Logs

$$\underline{\hspace{2cm}} = \underline{\hspace{3cm}}$$

The Log of a Power Equals the Product of the Power and the Log

$$\underline{\hspace{2cm}} = \underline{\hspace{2cm}}$$

3. **Example:** Write $\log_3\left[(x-1)(x+2)^2\right]$ as a sum of logarithms. Express all powers as factors.

Final answer: $\log_3\left[(x-1)(x+2)^2\right] = \underline{\hspace{4cm}}$

Note: After he completes the problem, he makes a comment about trying to rewrite $\log_3(x-1)$. This is a common mistake.

4. **Example:** Write $\log_5\left(\dfrac{x^2 y^3}{\sqrt{z}}\right)$ as a sum or difference of logs. Express all powers as factors.

Final answer: $\log_5\left(\dfrac{x^2 y^3}{\sqrt{z}}\right) = \underline{\hspace{4cm}}$

Course: Name:
Instructor: Section:

Section 11.4 – Objective 3: Write a Logarithmic Expression as a Single Logarithm
Video Length – 6:07

5. **Example:** Write each of the following as a single logarithm.

   (a) $\log_2 x + \log_2(x-3)$

   Final answer: $\log_2 x + \log_2(x-3) = $ _____

   (b) $3\log_6 z - 2\log_6 y$

   Final answer: $3\log_6 z - 2\log_6 y = $ _____

   (c) $\ln(x-2) + \dfrac{1}{2}\ln x - 5\ln(x+3)$

   Final answer: $\ln(x-2) + \dfrac{1}{2}\ln x - 5\ln(x+3) = $ _____

Course:
Instructor:
Name:
Section:

Section 11.4 – Objective 4: Evaluate a Logarithm Whose Base Is Neither 10 Nor e
Video Length – 9:38

6. **Example:** Approximate $\log_3 12$. Round answer to four decimal places.

   Final answer: $\log_3 12 \approx$ _____

   Change-of-Base Formula
   If $a \neq 1$, $b \neq 1$, and M are positive real numbers, then

   _____ = _____

   _____ = _____   and   _____ = _____

7. **Example:** Approximate $\log_7 325$.

   Final answer: $\log_7 325 \approx$ _____

   Note: Observe how he checks the reasonableness of his answer. This is an important skill to have when working with logarithms, or problem solving in general.

Course: Name:
Instructor: Section:

Section 11.5 Video Guide
Exponential and Logarithmic Equations

Objectives:
1. Solve Logarithmic Equations Using the Properties of Logarithms
2. Solve Exponential Equations
3. Solve Equations Involving Exponential Models

Section 11.5 – Objective 1: Solve Logarithmic Equations Using the Properties of Logarithms
Video Length – 9:16

Up to this point, we have been able to solve exponential equations by using the following:

Also, we solved logarithmic equations when the equation contained a single logarithm. We did this by converting the logarithm into an equivalent exponential equation. In other words, we used the following:

What if we have multiple logarithms in the equation? If you wish to solve a logarithmic equation with multiple logs, you can do one of two things.

1. If the equation is of the form _____ = _____, then ___ = ___ . In other words, if you have two logs with the same base equal to each other, then the arguments must be equal.

2. Use properties of logs to rewrite the equation with a single logarithm.

1. **Example:** Solve: $\log_3 4 = 2\log_3 x$

Final answer: _____
Note: Don't forget to check your answers!!!

Course: Name:
Instructor: Section:

2. **Example:** Solve: $\log_2(x+2) + \log_2(1-x) = 1$

Final answer: _____

Course: Name:
Instructor: Section:

Section 11.5 – Objective 2: Solve Exponential Equations
Video Length – 12:45

Up to this point, we have only solved exponential equations where each side of the equation can be written to the same base. Once written in this form, we solve the equation by setting the exponents equal to each other. Certainly, not all exponential equations can be solved this way.

3. **Example:** Solve: $3^x = 7$

   Final answer: _____

4. **Example:** Solve: $5 \cdot 2^x = 3$

   Final answer: _____

5. **Example:** Solve: $2^{x-1} = 5^{2x+3}$

   Final answer: _____

Course:
Instructor:

Name:
Section:

Section 11.5 – Objective 3: Solve Equations Involving Exponential Models
Video Length – 6:37

Compound Interest Formula
The future value of P dollars invested in an account paying an annual interest rate r, compounded n times per year for t years, is given by

$$\_\_ = _____.$$

6. **Example:** Suppose that you deposit $5000 into a Certificate of Deposit (CD) today. If the deposit earns 6% interest compounded monthly, how long will it be before the account is worth...

   (a) ...$7000?

   Final answer: _____

   (b) ...$10,000? That is, how long will you have to wait until your money doubles?

   Final answer: _____

Course: Name:
Instructor: Section:

Section 12.1 Video Guide
Distance and Midpoint Formulas

Objectives:
1. Use the Distance Formula
2. Use the Midpoint Formula

Section 12.1 – Objective 1: Use the Distance Formula
Video Length – 8:27

1. **Example:** Determine the distance between $(-3,-5)$ and $(3,3)$.

   Final answer: _____

The Distance Formula
The distance between two points $P_1 = (x_1, y_1)$ and $P_2 = (x_2, y_2)$, is

$$\rule{2cm}{0.4pt} = \rule{5cm}{0.4pt}$$

2. **Example:** Find the distance between $(-6,-6)$ and $(-5,-2)$.

   Final answer: _____

Copyright © 2014 Pearson Education, Inc.

Course:
Instructor:
Name:
Section:

Section 12.1 – Objective 2: Use the Midpoint Formula
Video Length – 2:43

Definition
The _____ of a line segment is the point located exactly halfway between the two endpoints of the line segment.

The Midpoint Formula
The midpoint $M = (x, y)$ of the line segment from $P_1 = (x_1, y_1)$ to $P_2 = (x_2, y_2)$ is

$$M = \underline{\hspace{2cm}}.$$

3. **Example:** Find the midpoint of the line segment joining $P_1 = (0, 8)$ and $P_2 = (4, -6)$.

Final answer: _____

Course: Name:
Instructor: Section:

Section 12.2 Video Guide
Circles

Objectives:
1. Write the Standard Form of the Equation of a Circle
2. Graph a Circle
3. Find the Center and Radius of a Circle Given an Equation in General Form

Section 12.2 – Objective 1: Write the Standard Form of the Equation of a Circle
Video Length – 5:44

Definition
A _____ is a set of all points in the Cartesian plane that are a fixed distance r from a fixed point (h,k). The fixed distance r is called the _____, and the fixed point (h,k) is called the _____ of the circle.

Definition
The _____ _____ of an equation of a circle with radius r and center (h,k) is

$$\underline{\hspace{3cm}} = \underline{\hspace{1cm}}.$$

1. **Example:** Determine the equation of the circle with radius 4 and center $(-5,2)$.

   Final answer: _____

Course:
Instructor:

Name:
Section:

Section 12.2 – Objective 2: Graph a Circle
Video Length – 4:07

2. **Example:** Graph the equation $(x+5)^2 + (y-2)^2 = 16$.

Course: Name:
Instructor: Section:

Section 12.2 – Objective 3: Find the Center and Radius of a Circle Given an Equation in General Form
Video Length – 7:28

Definition
The _____ _____ **of the equation of a circle** is given by the equation

$$\text{_____} = \text{\_\_\_\_}$$

when the graph exists.

3. **Example:** Graph the equation of the circle: $x^2 + y^2 + 2x - 8y + 8 = 0$

Course: Name:
Instructor: Section:

Section 12.3 Video Guide
Parabolas

Objectives:
1. Graph Parabolas in Which the Vertex Is the Origin
2. Find the Equation of a Parabola
3. Graph a Parabola Whose Vertex Is Not the Origin
4. Solve Applied Problems Involving Parabolas

Section 12.3 – Objective 1: Graph Parabolas in Which the Vertex Is the Origin
Video Length – 5:54

Definition
A _____ is defined as the collection of all points P in the plane that are the same distance from a fixed point F as they are from a fixed line D. The point F is called the _____ of the parabola, and the line D is its _____. As a result, a parabola is the set of points P for which

_____ = _____

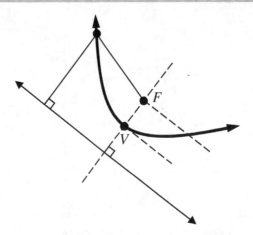

Now let's look at all four forms of a parabola whose vertex is at the origin.

Equation	Vertex	Focus	Directrix

Course:
Instructor:

Name:
Section:

Equation	Vertex	Focus	Directrix

Equation	Vertex	Focus	Directrix

Equation	Vertex	Focus	Directrix

1. **Example:** Graph the parabola $x^2 = -12y$.

Course: Name:
Instructor: Section:

Section 12.3 – Objective 2: Find the Equation of a Parabola
Video Length – 5:56

2. **Example:** Find an equation of the parabola with vertex at $(0,0)$ and focus at $(5,0)$. Graph the equation.

   Equation: _____

3. **Example:** Find the equation of a parabola with vertex at $(0,0)$ if its axis of symmetry is the y-axis and its graph contains the point $(2,3)$. Graph the equation.

   Equation: _____

Course: Name:
Instructor: Section:

Section 12.3 – Objective 3: Graph a Parabola Whose Vertex Is Not the Origin
Video Length – 6:32

Parabola with Axis of Symmetry Parallel to *x*-Axis, Opens to the Right, $a > 0$

Equation	Vertex	Focus	Directrix

Parabola with Axis of Symmetry Parallel to *x*-Axis, Opens to the Left, $a > 0$

Equation	Vertex	Focus	Directrix

Parabola with Axis of Symmetry Parallel to *y*-Axis, Opens Up, $a > 0$

Equation	Vertex	Focus	Directrix

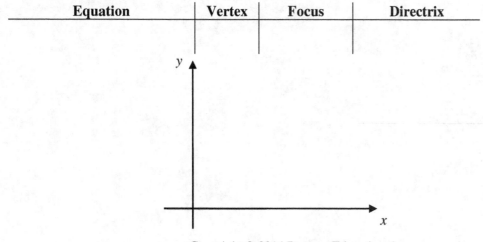

Course: Name:
Instructor: Section:

Parabola with Axis of Symmetry Parallel to y-Axis, Opens Down, $a > 0$

Equation	Vertex	Focus	Directrix

4. **Example:** Find an equation of the parabola with vertex at $(-2, 3)$ and focus at $(0, 3)$. Graph the equation.

Equation: _____

Course:
Instructor:
Name:
Section:

Section 12.3 – Objective 4: Solve Applied Problems Involving Parabolas
Video Length – 2:37

5. **Example:** A satellite dish is shaped like a paraboloid of revolution. The signals that emanate from a satellite strike the surface of the dish and are reflected to a single point, where the receiver is located. If the dish is 10 feet across at its opening and 4 feet deep at its center, at what position should the receiver be placed?

Final answer: _____

Note: Write your answer in a complete sentence.

Course: Name:
Instructor: Section:

Section 12.4 Video Guide
Ellipses

Objectives:
1. Graph an Ellipse Whose Center Is the Origin
2. Find the Equation of an Ellipse Whose Center Is the Origin
3. Graph an Ellipse Whose Center Is Not the Origin
4. Solve Applied Problems Involving Ellipses

Section 12.4 – Objective 1: Graph an Ellipse Whose Center Is the Origin
Video Length – 14:41

Definition
An _____ is the collection of all points in the plane the _____ of whose distances from two fixed points, called the _____, is a constant.

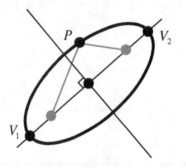

Equation of an Ellipse: Center at $(0,0)$; Major Axis along the x-Axis
An equation of the ellipse with center $(0,0)$, foci at $(-c,0)$ and $(c,0)$, and vertices at $(-a,0)$ and $(a,0)$ is

_____ = ____, where _____ and _____.

The major axis is the x-axis.

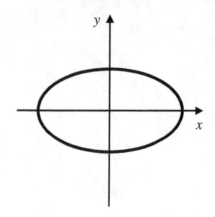

438   Copyright © 2014 Pearson Education, Inc.

Course:
Instructor:

Name:
Section:

1. **Example:** Analyze the equation: $\dfrac{x^2}{16} + \dfrac{y^2}{8} = 1$

Major axis: _____

Center: _____

Foci: _____

Vertices: _____

Equation of an Ellipse: Center at $(0,0)$; Major Axis along the y-Axis

An equation of the ellipse with center $(0,0)$, foci at $(0,-c)$ and $(0,c)$, and vertices at $(0,-a)$ and $(0,a)$ is

_____ = _____, where _____ and _____ .

The major axis is the y-axis.

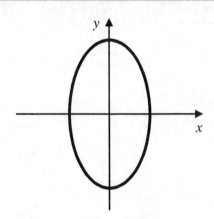

Course: Name:
Instructor: Section:

2. **Example:** Analyze the equation: $9x^2 + 4y^2 = 36$

Major axis: _____

Center: _____

Foci: _____

Vertices: _____

Course: Name:
Instructor: Section:

Section 12.4 – Objective 2: Find the Equation of an Ellipse Whose Center Is the Origin
Video Length – 6:12

3. **Example:** Find an equation of the ellipse with center at the origin, one focus at $(-3,0)$ and a vertex at $(5,0)$. Graph the equation.

Equation: _____

Course:
Instructor:

Name:
Section:

Section 12.4 – Objective 3: Graph an Ellipse Whose Center Is Not the Origin
Video Length – 12:54

4. **Example:** Analyze the equation: $25x^2 + 4y^2 + 150x + 16y + 91 = 0$

Major axis: _____

Center: _____

Foci: _____

Vertices: _____

Course: Name:
Instructor: Section:

Section 12.4 – Objective 4: Solve Applied Problems Involving Ellipses
Video Length – 3:25

Ellipses have an interesting reflection property. If a source of sound or light is placed at one focus, the waves transmitted by the source reflect off the ellipse and concentrate at the other focus. This is the principle behind whispering galleries, which are rooms with elliptical ceilings. A person standing at one focus of the ellipse can whisper and be heard by a person standing at the other focus, because all the sound waves that reach the ceiling are reflected to the other person.

5. **Example:** A whispering gallery is 50 feet long. The distance from the center of the room to the foci is 15 feet.

   (a) Find an equation that describes the shape of the room.

   Equation: _____

   (b) How high is the room at its center?

   Final answer: _____

Course:
Instructor:
Name:
Section:

Section 12.5 Video Guide
Hyperbolas

Objectives:
1. Graph a Hyperbola Whose Center Is the Origin
2. Find the Equation of a Hyperbola Whose Center Is the Origin
3. Find the Asymptotes of a Hyperbola Whose Center Is the Origin

Section 12.5 – Objective 1: Graph a Hyperbola Whose Center Is the Origin
Video Length – 19:37

Definition
A _____ is the collection of all points in the plane the _____ of whose distances from two fixed points, called the _____, is a constant.

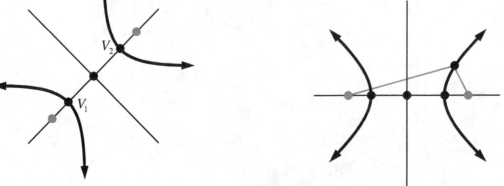

$$d(F_1, P) - d(F_2, P) = \pm 2a$$

Equation of a Hyperbola: Center at $(0,0)$; Transverse Axis along the x-Axis

An equation of the hyperbola with center $(0,0)$, foci at $(-c,0)$ and $(c,0)$, and vertices at $(-a,0)$ and $(a,0)$ is

_____ = _____, where _____.

The transverse axis is the x-axis.

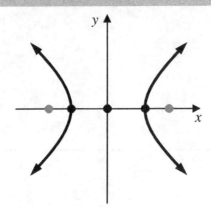

444 Copyright © 2014 Pearson Education, Inc.

Course: Name:
Instructor: Section:

1. **Example:** Analyze the equation $\dfrac{x^2}{25} - \dfrac{y^2}{16} = 1$

Transverse axis: _____

Center: _____

Foci: _____

Vertices: _____

Equation of a Hyperbola: Center at $(0,0)$; Transverse Axis along the y-Axis

An equation of the hyperbola with center $(0,0)$, foci at $(0,-c)$ and $(0,c)$, and vertices at $(0,-a)$ and $(0,a)$ is

$$\underline{\hspace{2in}} = \underline{\hspace{0.5in}}, \text{ where } \underline{\hspace{1.5in}}.$$

The transverse axis is the y-axis.

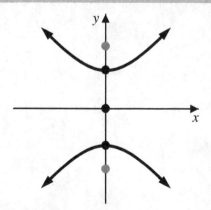

Course:
Instructor:

Name:
Section:

2. **Example:** Analyze the equation $2y^2 - 8x^2 = 32$

   Transverse axis: _____

   Center: _____

   Foci: _____

   Vertices: _____

Section 12.5 – Objective 2: Find the Equation of a Hyperbola Whose Center Is the Origin
Video Length – 9:03

3. **Example:** Find an equation of the hyperbola with center at the origin, one focus at $(-5,0)$, and one vertex at $(2,0)$. Graph the equation.

   Equation: _____

Course: Name:
Instructor: Section:

Section 12.5 – Objective 3: Find the Asymptotes of a Hyperbola Whose Center Is the Origin
Video Length – 11:31

Hyperbolas have a feature that circles and ellipses do not have; namely asymptotes.

Asymptotes of a Hyperbola

The hyperbola $\dfrac{x^2}{a^2} - \dfrac{y^2}{b^2} = 1$ has two oblique asymptotes

___ = ___   and   ___ = ___

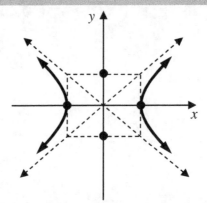

Asymptotes of a Hyperbola

The hyperbola $\dfrac{y^2}{a^2} - \dfrac{x^2}{b^2} = 1$ has two oblique asymptotes

___ = ___   and   ___ = ___

4. **Example:** Analyze the equation $y^2 - \dfrac{x^2}{16} = 1$

Transverse axis: _____

Center: _____

Foci: _____

Vertices: _____

Course: Name:
Instructor: Section:

Section 12.6 Video Guide
Systems of Nonlinear Equations

Objectives:
1. Solve a System of Nonlinear Equations Using Substitution
2. Solve a System of Nonlinear Equations Using Elimination

Section 12.6 – Objective 1: Solve a System of Nonlinear Equations Using Substitution
Video Length – 3:06

We are now going to solve systems of nonlinear equations. We can use the same methods that we learned when solving systems of linear equations. That is, we can use the method of substitution or elimination.

1. **Example:** Solve: $\begin{cases} 3x - y = -2 \\ 2x^2 - y = 0 \end{cases}$

Final answer: _____
Note: You heard him! Verify your solutions.

Course: Name:
Instructor: Section:

Section 12.6 – Objective 2: Solve a System of Nonlinear Equations Using Elimination
Video Length – 2:24

2. **Example:** Solve: $\begin{cases} x^2 + y^2 = 13 \\ x^2 - y = 7 \end{cases}$

Final answer: _____

Note: You can quickly verify that there are four solutions by graphing the equations on the same set of axes. However, you should always verify your work algebraically to make sure you didn't make any bonehead mistakes.

Course: Name:
Instructor: Section:

Section 13.1 Video Guide
Sequences

Objectives:
1. Write the First Few Terms of a Sequence
2. Find a Formula for the nth Term of a Sequence
3. Use Summation Notation

Section 13.1 – Objective 1: Write the First Few Terms of a Sequence
Video Length – 6:00

Definition
A _____ is a function whose domain is the set of positive integers.

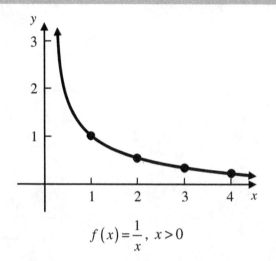

$f(x) = \dfrac{1}{x}, \ x > 0$

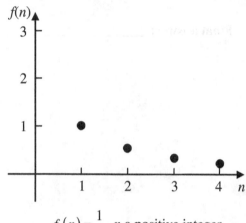

$f(n) = \dfrac{1}{n}, \ n$ a positive integer

We will use the following notation:

$$\{a_n\} = \left\{\dfrac{n-1}{n}\right\}$$

1. **Example:** Write down the first four terms of the following sequence.

$$\{a_n\} = \left\{\dfrac{n^2}{2n+1}\right\}$$

Final answer: _____

Copyright © 2014 Pearson Education, Inc. 451

Course: Name:
Instructor: Section:

2. **Example:** Write down the first five terms of the following sequence.

$$\{b_n\} = \{(-1)^n \cdot (2n)\}$$

Final answer: _____

Course: Name:
Instructor: Section:

Section 13.1 – Objective 2: Find a Formula for the nth Term of a Sequence
Video Length – 3:08

3. **Example:** Find a formula for the nth term of each sequence.

(a) $\dfrac{1}{2}, \dfrac{1}{4}, \dfrac{1}{8}, \dfrac{1}{16}, \ldots$

Final answer: _____

(b) 5, 7, 9, 11, …

Final answer: _____

Course:
Instructor:
Name:
Section:

Section 13.1 – Objective 3: Use Summation Notation
Video Length – 2:44

Summation notation is a shorthand way of representing the sum of a sequence of terms.

$$a_1 + a_2 + a_3 + \cdots + a_n = \sum_{k=1}^{n} a_k$$

This basically says to add up the first n terms of the sequence. For example

$$\sum_{k=1}^{8} k^2$$

4. **Example:** Express each sum using summation notation.

(a) $\left(\dfrac{1}{1}\right)^2 + \left(\dfrac{1}{2}\right)^2 + \left(\dfrac{1}{3}\right)^2 + \left(\dfrac{1}{4}\right)^2 + \left(\dfrac{1}{5}\right)^2$

Final answer: $\left(\dfrac{1}{1}\right)^2 + \left(\dfrac{1}{2}\right)^2 + \left(\dfrac{1}{3}\right)^2 + \left(\dfrac{1}{4}\right)^2 + \left(\dfrac{1}{5}\right)^2 =$ _____

Course: Name:
Instructor: Section:

Section 13.2 Video Guide
Arithmetic Sequences

Objectives:
1. Determine Whether a Sequence Is Arithmetic
2. Find a Formula for the *n*th Term of an Arithmetic Sequence
3. Find the Sum of an Arithmetic Sequence

Section 13.2 – Objective 1: Determine Whether a Sequence Is Arithmetic
Video Length – 8:39

In this section we are going to look at a specific type of sequence called an **arithmetic sequence**.

$$a_1 = a, \qquad a_n = a_{n-1} + d$$

The number d is referred to as the _____ _____.

Note: Observe the connection he makes with an arithmetic sequence and a linear function.

1. **Example:** Show that the sequence is arithmetic. List the first term and the common difference.

   (a) $4,\ 2,\ 0,\ -2,\ldots$

           Final answer: _____

   (b) $\{s_n\} = \{4n - 1\}$

           Final answer: _____

   (c) $\{t_n\} = \{2 - 3n\}$

           Final answer: _____

Course: Name:
Instructor: Section:

Section 13.2 – Objective 2: Find a Formula for the *n*th Term of an Arithmetic Sequence
Video Length – 11:32

Now we are going to find a general formula for the *n*th term of an arithmetic sequence. Ready?

nth Term of an Arithmetic Sequence
For an arithmetic sequence $\{a_n\}$ whose first term is a and whose common difference is d, the *n*th term is determined by the formula

_____ = _____

2. **Example:** Find the twenty fourth term of the arithmetic sequence:

$$-3, 0, 3, 6,\ldots$$

Final answer: _____

3. **Example:** The sixth term of an arithmetic sequence is 26, and the nineteenth term is 78.

   (a) Find the first term and the common difference.

   Final answer: _____

   (b) Give a formula for the *n*th term of the sequence.

   Final answer: _____

Course: Name:
Instructor: Section:

Section 13.2 – Objective 3: Find the Sum of an Arithmetic Sequence
Video Length – 8:38

> **Sum of n Terms of an Arithmetic Sequence**
> Let $\{a_n\}$ be an arithmetic with first term a and common difference is d. The sum S_n of the first n terms of $\{a_n\}$ is
>
> _____ = _____ = _____

4. **Example:** Find the sum of the first n terms of the sequence $\{4n+2\}$.

   Final answer: _____

5. **Example:** In the corner section of a theater, the first row has 20 seats. Each subsequent row has 2 more seats, and there are a total of 40 rows. How many seats are in this section?

   Final answer: _____

Course:
Instructor:
Name:
Section:

Section 13.3 Video Guide
Geometric Sequences and Series

Objectives:
1. Determine Whether a Sequence Is Geometric
2. Find a Formula for the nth Term of a Geometric Sequence
3. Find the Sum of a Geometric Sequence
4. Find the Sum of a Geometric Series
5. Solve Annuity Problems

Section 13.3 – Objective 1: Determine If a Sequence Is Geometric
Video Length – 6:39

A **geometric sequence** defined recursively is

$$a_1 = a, \qquad a_n = ra_{n-1}$$

The number r is referred to as the _____ _____.

1. **Example:** Show that the sequence is geometric. List the first term and the common ratio.

    (a) 2, 8, 32, 128,…

    Final answer: _____

    (b) $\{s_n\} = \{3^{n+1}\}$

    Final answer: _____

    (c) $\{t_n\} = \{3(2)^n\}$

    Final answer: _____

458   Copyright © 2014 Pearson Education, Inc.

Course: Name:
Instructor: Section:

Section 13.3 – Objective 2: Find a Formula for the *n*th Term of a Geometric Sequence
Video Length – 4:29

Now let's find a formula for a geometric sequence.

---

nth Term of a Geometric Sequence

For a geometric sequence $\{a_n\}$ whose first term is a and whose common ratio is r, the *n*th term is determined by the formula

$$\underline{} = \underline{}, \quad r \neq 0$$

---

2. **Example:** Find the ninth term of the geometric sequence:

    $$3,\ 2,\ \frac{4}{3},\ \frac{8}{9},\ \ldots$$

    Final answer: _____

Course: Name:
Instructor: Section:

Section 13.3 – Objective 3: Find the Sum of a Geometric Sequence
Video Length – 3:01

Sum of n Terms of a Geometric Sequence
Let $\{a_n\}$ be a geometric sequence with first term a and common ratio r, where $r \neq 0$, $r \neq 1$. The sum S_n of the first n terms of $\{a_n\}$ is

_____ = _____ , $r \neq 0, 1$

3. **Example:** Find the sum of the first n terms of the sequence $\{3^n\}$.

Final answer: _____

460

Course: Name:
Instructor: Section:

Section 13.3 – Objective 4: Find the Sum of a Geometric Series
Video Length – 8:32

Definition
An infinite sum of the form

$$a + ar + ar^2 + \cdots + ar^{n-1} + \cdots$$

with first term a and common ratio r, is called an _____ _____ _____ and is denoted by

4. **Example:** Find the sum of the geometric series: $1 + \dfrac{1}{3} + \dfrac{1}{9} + \cdots$

     Final answer: _____

5. **Example:** Express $0.8888\ldots$ as a fraction in lowest terms.

     Final answer: $0.8888\ldots =$ _____

Course: Name:
Instructor: Section:

Note: The following example is a detailed version of the one in the video.

6. **Example:** Suppose that, Americans spend 98% of every additional dollar they earn. Economists would say that an individual's **marginal propensity to consume** is 0.98. For example, if John earns $1, he will spend $0.98(\$1)$. Now suppose Jane earns $0.98(\$1)$ from John. She will then spend 98% of it or $0.98(0.98(\$1))$. This process of spending continues and results in a geometric series as follows:

$$\$1 + 0.98(\$1) + 0.98(0.98(\$1)) + 0.98(0.98(0.98(\$1))) + \cdots$$

Suppose the government gives a tax rebate of $1000 to John. Determine the impact of the rebate on the U.S. economy if Americans spend 98% of every dollar they earn.

Final answer: _____

Course: Name:
Instructor: Section:

Section 13.3 – Objective 5: Solve Annuity Problems
Video Length – 3:13

Amount of an Annuity

If P represents the deposit in dollars made at each payment period for an annuity at i percent interest per payment period, the amount A of the annuity after n payment periods is

$$\underline{} = \underline{}$$

7. **Example:** Christine contributes $100 each month into her 401(k) retirement plan. What will be the value of Christine's 401(k) in 30 years if the per annum rate of return is assumed to be 8% compounded monthly?

Final answer: _____

Course:　　　　　　　　　　　　　　　　　　　　　Name:
Instructor:　　　　　　　　　　　　　　　　　　　　Section:

Section 13.4 Video Guide
The Binomial Theorem

Objectives:
1. Compute Factorials
2. Evaluate a Binomial Coefficient
3. Expand a Binomial

Section 13.4 – Objective 1: Compute Factorials
Video Length – 1:51

The Factorial Symbol
If $n \geq 0$ is an integer, the **factorial symbol** _____ is defined as follows:

$0! =$ ___ $1! =$ ___

$n! =$ _____   if $n \geq 2$

For example:

Course: Name:
Instructor: Section:

Section 13.4 – Objective 2: Evaluate a Binomial Coefficient
Video Length – 2:17

Definition

If j and n are integers with $0 \le j \le n$, the symbol $\binom{n}{j}$ is defined as

$$\binom{n}{j} = \underline{\hspace{2cm}}$$

Note: The symbol $\binom{n}{j}$ is read as "n taken j at a time" or "n choose j."

1. **Example:** Find:

   (a) $\binom{4}{1}$

   Final answer: $\binom{4}{1} = \underline{\hspace{3cm}}$

   (b) $\binom{6}{2}$

   Final answer: $\binom{6}{2} = \underline{\hspace{3cm}}$

   (c) $\binom{8}{7}$

   Final answer: $\binom{8}{7} = \underline{\hspace{3cm}}$

Course:
Instructor:
Name:
Section:

Section 13.4 – Objective 3: Expand a Binomial
Video Length – 3:25

Binomial Theorem
Let x and a be real numbers. For any positive integer n, we have

$$(x+a)^n = \binom{n}{0}x^n + \binom{n}{1}ax^{n-1} + \cdots + \binom{n}{j}a^j x^{n-j} + \cdots + \binom{n}{n}a^n = \sum_{j=0}^{n} \binom{n}{j} a^j x^{n-j}$$

2. **Example:** Use the Binomial Theorem to expand $(x+2)^5$.

Final answer: $(x+2)^5 = $ _____

Course: Name:
Instructor: Section:

Appendix A Video Guide
Synthetic Division

Objectives:
1. Divide Polynomials Using Synthetic Division
2. Use the Remainder and Factor Theorems

Appendix A – Objective 1: Divide Polynomials Using Synthetic Division
Video Length – 10:41

Another approach that can be used to divide a polynomial by a polynomial is called **synthetic division**. HOWEVER, synthetic division CAN ONLY BE USED when the divisor is of the form $x-c$. For example,

$x-5$

1. **Example:** Use synthetic division: $(3x^2 + 2x - 1) \div (x - 2)$

   Final answer: $\dfrac{3x^2 + 2x - 1}{x - 2} = $ _____

2. **Example:** Divide by synthetic division: $(x^2 + 3x - 2) \div (x + 1)$

   Final answer: $\dfrac{x^2 + 3x - 2}{x + 1} = $ _____

Course: Name:
Instructor: Section:

Appendix A – Objective 2: Use the Remainder and Factor Theorems
Part I – Text Example 3
Video Length – 5:34

The Remainder Theorem
Let f be a polynomial function. If $f(x)$ is divided by _____ , then the remainder is _____ .

3. **Example:** Use the Remainder Theorem to find the remainder of $f(x) = x^2 + 4x - 5$ is divided by $x + 2$.

   Final answer: _____

468

Course: Name:
Instructor: Section:

Appendix A – Objective 2: Use the Remainder and Factor Theorems
Part II – Text Example 4
Video Length – 11:54

The Factor Theorem
Let f be a polynomial function. Then _____ is a factor of $f(x)$ if and only if _____ .

4. **Example:** Use the Factor Theorem to determine whether the function $f(x) = 3x^2 + x - 2$ has a factor of $x + 2$.

   Final answer: _____

5. **Example:** Consider the function $f(x) = 2x^3 - x^2 - 16x + 15$.

   (a) Is $x - 2$ a factor of f?

   Final answer: _____

   Use synthetic division to verify your answer above.

   (b) Is $x + 3$ a factor of f?

   Final answer: _____

   Use synthetic division to verify your answer above.

Course:
Instructor:

Name:
Section:

Course: Name:
Instructor: Section:

Appendix C.1 Video Guide
A Review of Systems of Linear Equations in Two Variables

Objectives:
1. Determine Whether an Ordered Pair Is a Solution of a System of Linear Equations
2. Solve a System of Two Linear Equations by Graphing
3. Solve a System of Two Linear Equations by Substitution
4. Solve a System of Two Linear Equations by Elimination
5. Identify Inconsistent Systems
6. Write the Solution of a System with Dependent Equations

Appendix C.1 – Objective 1: Determine Whether an Ordered Pair Is a Solution of a System of Linear Equations
Video Length – 6:14

Recall that a linear equation is an equation that can be written in the form $Ax + By = C$, where A, B and C are real numbers and A and B cannot simultaneously be zero. This is a linear equation in two variables. We don't necessarily have to limit ourselves to linear equations in two variables. We can also talk about linear equations in three variables. Or we can talk about linear equations in four variables.

Definition
A _____ of _____ _____ is a grouping of two or more linear equations, each of which contains one or more variables.

Examples of systems of linear equations:

Definition
A _____ of a system of equations consists of values for the variables that are solutions of each equation of the system.

1. **Example:** Determine whether $(-4, 16)$ is a solution to the system of the equations.

$$\begin{cases} y = -4x \\ y = -2x + 8 \end{cases}$$

Final answer: _____

Is $(-2, 8)$ a solution?

Course: Name:
Instructor: Section:

Graph	Number of Solutions	Type of System
Two lines _____ at one point.	If the lines _____, the system of equations has _____ solution given by the _____ of _____ .	_____ The equations are _____
_____ lines	If the lines are _____ , then the system of equations has _____ solution because the lines _____ _____ .	_____
Lines _____	If the lines _____ on top of each other, then the system has _____ many solutions. The solution set is the set of _____ _____ on the _____.	_____ The equations are _____ .

Course: Name:
Instructor: Section:

Appendix C.1 – Objective 2: Solve a System of Two Linear Equations by Graphing
Video Length – 5:06

Now let's solve a system of linear equations containing two unknowns by graphing.

2. **Example:** Solve the following system by graphing. $\begin{cases} 3x + y = -6 \\ 2x - y = 1 \end{cases}$

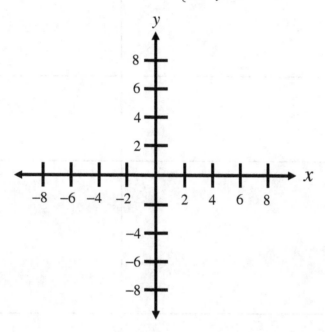

Final answer: _____

Course:
Instructor:
Name:
Section:

Appendix C.1 – Objective 3: Solve a System of Two Linear Equations by Substitution
Video Length – 8:31

3. **Example:** Solve the following system by substitution: $\begin{cases} 2x - y = 13 \\ -4x - 9y = 7 \end{cases}$

Write the steps in words	Show the steps with math
Step 1	
Step 2	
Step 3	
Step 4	
Step 5 CHECK YOUR WORK!!!	

Final answer: _____

Course: Name:
Instructor: Section:

Appendix C.1 – Objective 4: Solve a System of Two Linear Equations by Elimination
Video Length – 10:12

Note: Listen carefully to how he chooses which variable to eliminate.

4. **Example:** Solve the following system by elimination: $\begin{cases} \dfrac{3}{2}x - \dfrac{y}{8} = -1 \\ 16x + 3y = -28 \end{cases}$

Write the steps in words	Show the steps with math
Step 1	
Step 2	
Step 3	
Step 4 *Note: This is the step where he determines y.*	
Step 5 CHECK YOUR WORK!!!	

Final answer: _____

Course: Name:
Instructor: Section:

Appendix C.1 – Objective 5: Identify Inconsistent Systems
Video Length – 7:29

Now we are going to look at **inconsistent systems**.

5. **Example:** Solve by graphing: $\begin{cases} 3x - 2y = -4 \\ -9x + 6y = -1 \end{cases}$

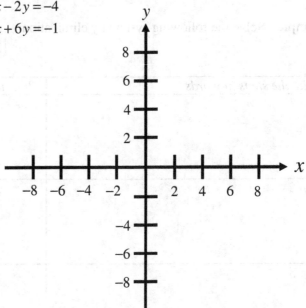

Final answer: _____

6. **Example:** Solve: $\begin{cases} 3x + 6y = 12 \\ x + 2y = 7 \end{cases}$

Final answer: _____

Course:
Instructor:

Name:
Section:

Appendix C.1 – Objective 6: Write the Solution of a System with Dependent Equations
Video Length – 7:41

Note: Get ready for some SWEET notation!

7. **Example:** Solve by graphing: $\begin{cases} 4x - 6y = 8 \\ -2x + 3y = -4 \end{cases}$

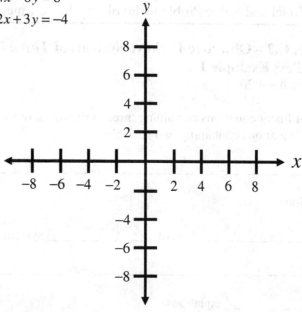

Final answer: _____

8. **Example:** Solve: $\begin{cases} 6x - 4y = 8 \\ -9x + 6y = -12 \end{cases}$

Final answer: _____

Copyright © 2014 Pearson Education, Inc.

Course: Name:
Instructor: Section:

Appendix C.2 Video Guide
Systems of Linear Equations in Three Variables

Objectives:
1. Solve Systems of Three Linear Equations
2. Identify Inconsistent Systems
3. Write the Solution of a System with Dependent Equations
4. Model and Solve Problems Involving Three Linear Equations

Appendix C.2 – Objective 1: Solve Systems of Three Linear Equations
Part I – Text Example 1
Video Length – 4:20

Systems of linear equations containing three variables have the same possible solutions as a system of two linear equations containing two variables:

1. _____ ____ _____ – A _____ system with _____ equations.

2. _____ _____ – An _____ system

3. _____ _____ _____ – A _____ system with _____ equations.

Recall that a _____ to a system of equations consists of values for the variables that are solutions of each equation of the system. The graph of each equation in a system of linear equations containing three variables is a plane in space.

1. **Example:** Determine if $(-3, 0, 1)$ is a solution to the system of equations.

$$\begin{cases} 2x + 5y - z = -7 \\ x - 11y + 4z = 1 \\ -5x + 8y - 12z = 3 \end{cases}$$

Final answer: _____

Course: Name:
Instructor: Section:

Appendix C.2 – Objective 1: Solve Systems of Three Linear Equations
Part II – Text Example 2
Video Length – 6:39

Typically when we solve a system of three equations containing three unknowns we use the elimination method. The basic idea is to use elimination in order to rewrite the system of three equations containing three unknowns to a system of two linear equations containing two unknowns. And then we can solve that system using the substitution or the elimination method.

2. **Example:** Use the elimination method to solve the system.

$$\begin{cases} 3x + 9y + 6z = 3 \\ 2x + y - z = 2 \\ x + y + z = 2 \end{cases}$$

Final answer: _____

Note: CHECK YOUR ANSWER!!! That is, make sure the ordered triple satisfies EACH of the three original equations in the system.

Course: Name:
Instructor: Section:

Appendix C.2 – Objective 2: Identify Inconsistent Systems
Video Length – 2:13

3. **Example:** Use the elimination method to solve the system.

$$\begin{cases} 3x + y - 2z = 2 \\ -6x - 2y + 4z = -2 \\ 9x + 3y - 6z = 6 \end{cases}$$

Final answer: _____

Course: Name:
Instructor: Section:

Appendix C.2 – Objective 3: Write the Solution of a System with Dependent Equations
Video Length – 5:43

4. **Example:** Use the elimination method to solve the system.

$$\begin{cases} x+y-2z=3 \\ -2x-3y+z=-7 \\ x+2y+z=4 \end{cases}$$

Final answer: _____

Course:
Instructor:

Name:
Section:

Appendix C.2 – Objective 4: Model and Solve Problems Involving Three Linear Equations

Video Length – 5:43

5. **Example:** A survey was conducted in an elementary school about favorite colors. A total of 235 students were surveyed. Three times as many students preferred the color blue as did students whose favorite color was green. Eleven more students like red better than blue. How many students liked each of the three colors?

Final answer: _____

Course: Name:
Instructor: Section:

Appendix C.3 Video Guide
Using Matrices to Solve Systems

Objectives:
1. Write the Augmented Matrix of a System
2. Write the System from the Augmented Matrix
3. Perform Row Operations on a Matrix
4. Solve Systems Using Matrices
5. Solve Consistent Systems with Dependent Equations and Inconsistent Systems

Appendix C.3 – Objective 1: Write the Augmented Matrix of a System
Video Length – 5:59

Definition
A _____ is defined as a rectangular array of numbers.

Examples:

A matrix has rows and columns. The number of rows and columns are used to "name" the matrix.

The plural of matrix is "**matrices**."

Definition
_____ _____ are used to represent systems of linear equations.

The system $\begin{cases} 2x+3y-z=12 \\ x-8y=16 \end{cases}$ is written as the augmented matrix

1. **Example:** Write the following system of equations as an augmented matrix.

$$\begin{cases} 3x+4y-2z=9 \\ -2x-y+z=0 \\ x-6z=3 \end{cases}$$

Final answer: _____

Course: Name:
Instructor: Section:

Appendix C.3 – Objective 2: Write the System from the Augmented Matrix
Video Length – 3:07

2. **Example:** Write the system of equations that corresponds to each augmented matrix.

(a) $\begin{bmatrix} 4 & -5 & | & 15 \\ 7 & 8 & | & -9 \end{bmatrix}$

(a) **Final answer:** _____

(b) $\begin{bmatrix} 2 & 0 & -6 & | & 3 \\ -8 & 7 & 3 & | & 12 \\ -5 & 8 & 4 & | & -3 \end{bmatrix}$

(b) **Final answer:** _____

Course: Name:
Instructor: Section:

Appendix C.3 – Objective 3: Perform Row Operations on a Matrix
Video Length – 11:58

Row operations are used on augmented matrix to solve the corresponding system of equations.

Note: Listen carefully to the analogous relationship between operations on a system of equations and the following row operations.

Row Operations
1. _____ any two rows.

2. Replace a row by a _____ _____ of that row.

3. Replace a row by the _____ of that row and a nonzero multiple of some other row.

Notation for row operations: (*Note: The following is for notational purposes only. For now, just understand the notation and how it corresponds to the row operations.*)

$$\begin{bmatrix} 4 & -5 & | & 15 \\ 7 & 8 & | & -9 \end{bmatrix}$$

Row _____ .

$$\begin{bmatrix} 4 & -5 & | & 15 \\ 7 & 8 & | & -9 \end{bmatrix}$$

Every _____ .

$$\begin{bmatrix} 4 & -5 & | & 15 \\ 7 & 8 & | & -9 \end{bmatrix}$$

Every _____ .

The main reason for using a matrix to solve a system of equations is that it is more efficient notation and helps us organize the mathematics.

Course: Name:
Instructor: Section:

3. **Example:** For the augmented matrix

$$\begin{bmatrix} 1 & -5 & | & 8 \\ 0 & 1 & | & 4 \end{bmatrix}$$

find a row operation that would result in the entry in row 1, column 2 become a 0.

Final answer: _____

Course: Name:
Instructor: Section:

Appendix C.3 – Objective 4: Solve Systems Using Matrices
Part I – Text Example 5
Video Length – 16:31

Definition
A matrix is in _____ _____ _____ when

1. The entry in row 1, column 1, is a _____ and _____ appear below it.

2. The first _____ entry in each row after the first row is a _____, _____ appear below it, and it appears to the _____ of the first nonzero entry in any row _____ .

3. Any rows that contain all 0s to the left of the vertical bar appear at the _____ .

Example:

3. **Example:** Solve: $\begin{cases} 3x + 11y = 13 \\ x + 5y = 7 \end{cases}$

Final answer: _____
Note: CHECK YOUR ANSWER!!!

Course: Name:
Instructor: Section:

Appendix C.3 – Objective 4: Solve Systems Using Matrices
Part II – Text Example 6
Video Length – 16:00

4. **Example:** Solve: $\begin{cases} x+y-z=-1 \\ 4x-3y+2z=16 \\ 2x-2y-3z=5 \end{cases}$

Final answer: _____
Note: CHECK YOUR ANSWER!!!

Course:
Instructor:

Name:
Section:

Appendix C.3 – Objective 5: Solve Consistent Systems with Dependent Equations and Inconsistent Systems
Video Length – 7:22

5. **Example:** Solve: $\begin{cases} -6x + 12y + 3z = -6 \\ 3x + 4y - z = 5 \\ -x + 2y + \frac{1}{2}z = -1 \end{cases}$

Final answer: _____

Course:
Instructor:

Name:
Section:

Appendix C.4 Video Guide
Determinants and Cramer's Rule

Objectives:
1. Evaluate the Determinant of a 2×2 Matrix
2. Use Cramer's Rule to Solve a System of Two Equations
3. Evaluate the Determinant of a 3×3 Matrix
4. Use Cramer's Rule to Solve a System of Three Equations

Appendix C.4 – Objective 1: Evaluate the Determinant of a 2×2 Matrix
Video Length – 2:07

Definition
If a, b, c, and d are four real numbers, the symbol

$$D =$$

is called a ___ ___ ___ _____. Its value is the number _____; that is

$$D =$$

1. **Example:** Evaluate: $\begin{vmatrix} -2 & 3 \\ 4 & -1 \end{vmatrix}$

   Final answer: $\begin{vmatrix} -2 & 3 \\ 4 & -1 \end{vmatrix} =$ _____

Course: Name:
Instructor: Section:

Appendix C.4 – Objective 2: Use Cramer's Rule to Solve a System of Two Equations
Video Length – 6:52

Now that we know how to compute the value of 2 by 2 determinants, we can now use **Cramer's Rule** to solve a system of two linear equations containing two variables.

Cramer's Rule for Two Equations Containing Two Variables
The solution to the system of equations

$$\begin{cases} ax + by = s \\ cx + dy = t \end{cases}$$

is given by

$$x = \underline{}, \qquad y = \underline{}$$

provided that

$$D = \begin{vmatrix} a & b \\ c & d \end{vmatrix} = ad - bc \neq \underline{}.$$

We can also look at Cramer's Rule in the following way.

Given the system $\begin{cases} ax + by = s \\ cx + dy = t \end{cases}$ with $D = \begin{vmatrix} a & b \\ c & d \end{vmatrix}$, define

$$D_x = \qquad\qquad D_y =$$

If $D \neq 0$, then

$$x = \frac{}{}, \qquad y = \frac{}{}$$

2. **Example:** Use Cramer's Rule, if applicable, to solve the system

$$\begin{cases} 3x - 6y = 24 \\ 5x + 4y = 12 \end{cases}$$

Final answer: _____

Note: Remember, don't make any bonehead mistakes!!!

Course: Name:
Instructor: Section:

Appendix C.4 – Objective 3: Evaluate the Determinant of a 3×3 Matrix
Video Length – 5:55

Note: Make a note of the meaning of the subscripts used in the following definition.

Definition
A ___ ___ ___ _____ is symbolized by

$$\begin{vmatrix} a_{11} & a_{12} & a_{13} \\ a_{21} & a_{22} & a_{23} \\ a_{31} & a_{32} & a_{33} \end{vmatrix}$$

in which $a_{11}, a_{12}, \ldots$, are real numbers.

$$\begin{vmatrix} a_{11} & a_{12} & a_{13} \\ a_{21} & a_{22} & a_{23} \\ a_{31} & a_{32} & a_{33} \end{vmatrix} =$$

3. **Example:** Find the value of the 3 by 3 determinant: $\begin{vmatrix} 1 & 2 & 1 \\ 3 & 5 & 1 \\ 2 & 6 & 7 \end{vmatrix}$

Final answer: _____

Course: Name:
Instructor: Section:

Appendix C.4 – Objective 4: Use Cramer's Rule to Solve a System of Three Equations
Video Length – 6:46

We can also use Cramer's Rule to solve systems of three equations.

Cramer's Rule for Three Equations Containing Three Variables
For the system of three equations containing three variables

$$\begin{cases} a_{11}x + a_{12}y + a_{13}z = c_1 \\ a_{21}x + a_{22}y + a_{23}z = c_2 \\ a_{31}x + a_{32}y + a_{33}z = c_3 \end{cases}$$

with $D = \begin{vmatrix} a_{11} & a_{12} & a_{13} \\ a_{21} & a_{22} & a_{23} \\ a_{31} & a_{32} & a_{33} \end{vmatrix} \neq 0$ $D_x =$ $D_y =$ $D_z =$

then

$$x = \underline{} \qquad y = \underline{} \qquad z = \underline{}$$

4. **Example:** Use Cramer's Rule, if applicable, to solve the following system:

$$\begin{cases} x + 2y + z = 1 \\ 3x + 5y + z = 3 \\ 2x + 6y + 7z = 1 \end{cases}$$

Final answer: _____